KINETICS OF HETEROGENEOUS CATALYTIC REACTIONS

PHYSICAL CHEMISTRY:
SCIENCE AND ENGINEERING

SERIES EDITORS:
John M. Prausnitz
Leo Brewer

KINETICS OF HETEROGENEOUS CATALYTIC REACTIONS

Michel Boudart and
G. Djéga-Mariadassou

Princeton University Press / Princeton, N.J.

Copyright © 1984 by Princeton University Press
Published by Princeton University Press,
41 William Street, Princeton, New Jersey 08540
In the United Kingdom: Princeton University Press,
Guildford, Surrey

All Rights Reserved
Library of Congress Cataloging in Publication Data
will be found on the last printed page of this book
ISBN cloth 0-691-08346-0   paper 0-691-08347-9
First published in French by Masson, S. A., 1982.

This book has been composed in Lasercomp Times Roman

Clothbound editions of Princeton University Press books
are printed on acid-free paper, and binding materials
are chosen for strength and durability.
Paperbacks, while satisfactory for personal collections,
are not usually suitable for library rebinding.

Printed in the United States of America by
Princeton University Press
Princeton, New Jersey

To Sir Hugh Taylor

TABLE OF CONTENTS

FIGURES

TABLES

PREFACE TO THE FRENCH EDITION

This book originated as a series of lectures given during the spring of 1980 at the Université Pierre et Marie Curie by Michel Boudart, at the request of Professor Jacques Fraissard, whose generous hospitality is gratefully acknowledged. As the lectures were being given, a close collaboration started to develop between the two authors of this book, who decided to write up the lectures. We believe this book is timely.

Indeed, interest in heterogeneous catalysis is growing rapidly among surface scientists and chemical engineers. This is because so many problems related to environment, energy, food, and materials are being solved by heterogeneous catalysis. As examples, let us cite the cleaning of automotive exhaust by a catalyst containing metals of the platinum family, the synthesis of high octane gasoline from methanol on a new zeolite, and the increasing use of steam reforming of methane on catalysts to produce hydrogen, either for ammonia synthesis toward the production of nitrogen fertilizers or for the reduction of iron ores for the manufacture of steel.

As a catalytic process moves from the laboratory to the pilot plant, and finally to the commercial unit, a quantitative kinetic treatment of the reaction is required. And in the laboratory where the process is developed, kinetics is the essential quantitative research tool to screen and optimize possible catalysts. This book is directed to scientists and engineers concerned with heterogeneous catalysis in academic teaching and research, as well as in industrial research and development.

M. Boudart, Stanford University
G. Djéga-Mariadassou, Université Pierre et Marie Curie
September 1981

INTRODUCTION

During its lifetime, any catalyst produces a large number of product molecules per molecule of active phase. This quantity is frequently called the *turnover*. The rate of the catalytic act, called the *activity* of the catalyst, is the *turnover frequency*, or the number of turnovers per second.

To understand the activity at the molecular level or to build a catalytic reactor on an industrial scale, the first necessary information is the turnover frequency and its variation with process variables, temperature, total pressure, and composition. This variation is described by a *rate equation*. To obtain the latter is an art based on the principles of chemical kinetics.

The reputation of chemical kinetics is unfortunately slightly tarnished, because kinetic data—necessary as they may be—will usually remain insufficient. They must be bolstered by spectroscopic data that define the composition and structure of the catalyst, and all the intermediates involved in the *elementary steps* that constitute the catalytic reaction.

The limitations of kinetics are particularly severe in heterogeneous catalysis. The traditional ignorance of the nature of catalytic surfaces has frequently led to the elaboration of rate equations, which seem to have only the merit of providing a practical representation of the data required for the design, optimization, and control of reactors. Thus, kinetics in heterogeneous catalysis is often considered as a tool of great use to the engineer but of little interest to the laboratory chemist.

This situation has been changed during the past fifteen years as a result of rapid progress in what is now known as *surface science*, founded on the physical examination of surfaces by a large and increasing number of new spectroscopic methods. Our knowledge on the structure and reactivity of surfaces has been particularly developed in the case of metals, which represent a very important class of catalysts. More specifically, research done on single crystals in ultrahigh vacuum chambers has led for the first time to the accumulation of kinetic data on catalytic elementary steps, namely *adsorption, surface reaction,* and *desorption*. Moreover, it is already clear that these kinetic results obtained at very low pressure on macroscopic crystals may be applicable qualitatively and sometimes even quantitatively to catalytic reactions taking place on microscopic clusters at normal pressure.

In parallel with the development of surface science, the overall kinetics of catalytic reactions on metals has recently become quantitative and reproducible. Indeed, for a number of reactions catalyzed by commercial catalysts we now know the turnover frequency under well-defined conditions. This turnover frequency has often been reproduced in different laboratories on catalysts prepared by different researchers. Thus, the myth of the art of catalysis protected by patents and trade secrets has been dissipated. The physical and chemical characterization of metallic catalysts can lead today to quantitative kinetics forming the basis of a new *catalysis science*.

It is therefore timely to write a critical account of catalytic kinetics, the growth of which is measured by developments in surface science and in catalysis science. This account can be written today for catalysis by metals, the subject of the present book.

Our goal is to present the principles of heterogeneous kinetics applied to catalytic metals. We are writing for the researcher who is interested in the synthesis and in the physicochemical study of the catalyst. Any study of this kind remains incomplete without a measure of the turnover frequency, which is the true evaluation of catalytic activity. We are also writing for the engineer in research, design, or plant operation who wishes to understand the rate equation that he uses today, but frequently wishes to improve in the future.

Kinetics is not only a tool of pure or applied research, but also a very satisfying avocation. Reaction mechanisms come and go, and their ephemeral existence is often disconcerting. By contrast, the results of good chemical kinetics remain unchanged, whatever may be the future revisions of their underlying mechanism. The chemist in chemical dynamics is frequently accused by the engineer of *explaining* everything without ever *predicting* anything. Yet the primary goal of kinetics is to *describe* the chemical transformation. A good description possesses permanent value, which remains the foundation of any future explanation or prediction. To contribute something of lasting value is a normal human aspiration. The kinetics of the pioneers of heterogeneous catalysis has retained all of its value. Today, after a temporary eclipse, heterogeneous catalytic kinetics has rejoined the vanguard of surface science and catalysis science, and has become once more a respectable endeavor for chemists and chemical engineers.

KINETICS OF HETEROGENEOUS CATALYTIC REACTIONS

Chapter 1

CONCEPTS AND DEFINITIONS

1.1 INTRODUCTION

Heterogeneous catalysis is much more than a subfield of chemical dynamics and chemical kinetics. It is related to other disciplines, as shown in the triangular representation below.

In particular, thanks to the recent development of the chemical physics of metallic surfaces, kineticists have reconsidered earlier views and theories concerning catalysis by metals and alloys. New techniques had yielded new results, and new concepts had to be incorporated in the kinetic framework of heterogeneous catalysis.

Catalyst preparation is responsible for the composition, structure and texture of catalytic materials. Today, the synthesis of new metallic catalysts is achieved in a more rational manner than heretofore, by means of solution theory, colloidal chemistry, solid state chemistry, and organometallic chemistry. At any rate, the most striking recent advance in catalysis by metals is the control of catalyst preparation and characterization, so that reproducible surfaces yield reproducible reactivity in different laboratories.

Besides helping in the characterization of catalytic solids, new physical tools (spectroscopy, diffraction) also contribute to the identification of *reaction intermediates* responsible for the elementary steps that constitute the catalytic cycle.

Ultimately, the indispensable quantity for the elaboration of theories in catalysis is the *correct* and *reproducible* measurement of the true reactivity of molecules at the solid-fluid interface. This measurement is the difficult art of the chemical kineticist. The measurement must be checked so that it can be shown to be exempt from all artifacts: adventitious poisoning, improper contacting between solid and fluid, and all physical processes of heat and mass transfer. But if the kinetic data are correct, and if the catalyst characterization is adequate, the available information can lead to the progressive transformation of catalysis from an art to a science.

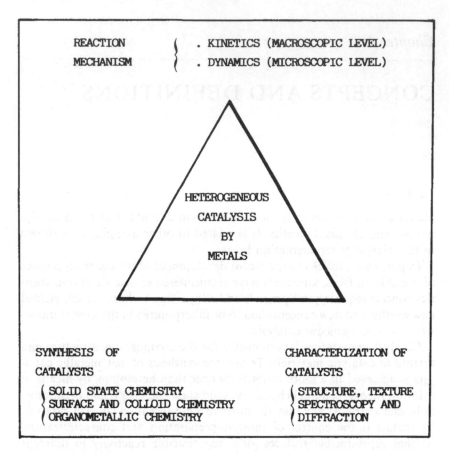

In fact, in the case of metals, surface science and catalysis science have progressed side by side in the past ten years. There now exist several examples where kinetic data on clean, well-defined macroscopic single crystals agree very well with those obtained on reproducible and characterized supported metallic clusters between 1 and 10 nm.

This remarkable agreement justifies the limitation of this book to *metallic catalysts*. Nevertheless, our attempt will be to develop the general kinetic principles involved in heterogeneous catalysis, with metals selected as an example.

The theories underlying these principles are relatively few. In spite of the post-Langmuirian era of surface science starting twenty years ago, the *Langmuir* (1916) *isotherm* remains one of the pillars that support surface catalytic kinetics. Nevertheless, it is now necessary to take into account *corrosive chemisorption* in which the adsorbate forms with the atoms of the metal surface a new coincidence lattice over the metal lattice itself (Bénard, 1970; Ponec and Sachtler, 1972; Hanson and Boudart, 1978). The theoretical developments of Horiuti (1957, 1967) are of wide applicability. But the two most essential theories of almost universal applicability remain that of the *transition state* or activated complex, elaborated by Eyring and his school (Glasstone et al., 1941) and the *quasi-stationary state* approximation popularized and defended by Bodenstein. Bodenstein's powerful method was further systematized by Christiansen (1953) for both catalytic and chain reactions.

In the first chapters of this book, the metal surface will be considered as made up of sites that are identical thermodynamically and kinetically without any interaction between adsorbed species. This Langmuirian view will be amended later as the early recognition by Taylor (1925) of the importance of active centers will be embodied in the theory of non-uniform surfaces following Temkin (1957, 1965, 1979) and Wagner (1970). Temkin's formalism, which rests on the Brønsted relation between rate constants and equilibrium constants (see Boudart, 1968), is very general. Since a collection of sites on a non-uniform surface can be assimilated to an array of different catalysts, Temkin's theory helps in understanding what determines the optimum catalyst for a given reaction. The finding of an optimum catalyst is the most common goal of applied catalysis.

1.2 DEFINITIONS

1.21 Stoichiometric Equation and Stoichiometric Coefficients

Generally, for any chemical reaction, whether it be an overall reaction or an elementary step, we can write:

$$0 = \sum_i v_i B_i \qquad (1.2.1)$$

where v_i is the stoichiometric coefficient of component B_i, taken as positive if B_i is a product, or negative if B_i is a reactant.

1.22 Extent of Reaction

This quantity, introduced by the Brussels school of thermodynamics, is defined by:

$$\xi \, (\text{mol}) = \frac{n_i - n_i^o}{v_i} \qquad (1.2.2)$$

where n_i^o and n_i are the quantity of substance of B_i, expressed in mole, at time zero or at any time respectively.

1.23 Reaction Rate

Following the recommendations of IUPAC (1976, 1979), reaction rate is defined in the most general way by:

$$\xi = \frac{d\xi}{dt} \text{ mol s}^{-1} \qquad (1.2.3)$$

But in practice, the rate is referred to unit volume, mass or area of the catalyst. Thus we can have a *volumic rate*:

$$\frac{1}{V}\frac{d\xi}{dt} \text{ mol cm}^{-3} \text{ s}^{-1} \qquad (1.2.4)$$

where V is the volume of the solid catalyst, or a *specific rate*:

$$\frac{1}{m}\frac{d\xi}{dt} \text{ mol g}^{-1} \text{ s}^{-1} \qquad (1.2.5)$$

where m is the mass of the catalyst, or an *areal rate*:

$$\frac{1}{A}\frac{d\xi}{dt} \text{ mol cm}^{-2} \text{ s}^{-1} \qquad (1.2.6)$$

where A is the area of the catalyst. These expressions for the rate of reaction will be designated by v, the meaning of which will be made clear by the context.

It is clear that the areal rate (1.2.6) is the best one of the three. Yet, two catalysts can have the same surface area but different concentrations of active sites. A definition of the rate in terms of the number of active sites would seem to be preferred.

1.24 Number of Turnovers

This is the number n of times that the overall reaction takes place through the catalytic cycle. The rate is then:

$$\text{rate} = \frac{dn}{dt} \text{ s}^{-1} \qquad (1.2.7)$$

where $n = \xi \times N_A$ with $N_A = 6.0225 \times 10^{23}$ mol^{-1}. The areal rate can also be expressed as:

$$\text{areal rate} = \frac{1}{A}\frac{dn}{dt} \text{ cm}^{-2}\text{ s}^{-1} \qquad (1.2.8)$$

1.25 Turnover Frequency or Rate of Turnover
(Formerly Called Turnover Number, Boudart 1972)

This is the number of turnovers per catalytic site and per unit time, for a reaction at a given temperature, pressure, reactant ratio, and extent of reaction. The latter may be so small that the rate may be considered as initial rate.

The turnover frequency is thus:

$$v_t \equiv N(s^{-1}) = \frac{1}{S}\frac{dn}{dt} \qquad (1.2.9)$$

where S is the number of active sites used in the experiment. Also:

$$S = [L] \times A \qquad (1.2.10)$$

where $[L]$ is the number density (cm^{-2}) of sites, i.e., the number of sites per unit area. The definition (1.2.9) will be used whenever possible. Not the least of its merit is that its dimension is one over time so that it is simply expressed in reciprocal seconds. This permits an easy comparison between the work of different investigators.

But there remain problems in the use of v_t. Indeed, we still do not know how to count the number of active sites. With metals, all we can do is count the number of exposed surface atoms. How many of the latter are grouped in an active site? Besides, different types of active sites probably always exist, and a molecule may be adsorbed differently on each type and react at a different rate. Thus v_t is likely to be an *average* value of the catalytic *activity* and a lower bound to the true activity, as only a fraction of the surface atoms may contribute to the activity. Besides, v_t is a rate, not a rate constant, so that it is always necessary to specify all conditions of reaction as spelled out above.

Yet, the systematic use of the turnover frequency is an immense progress made possible by techniques that count the number of surface metallic atoms. It is likely that the progress made in the case of metals will also be made with oxides (Djéga Mariadassou et al., 1982), in particular, the comparison between the activity of large single crystals and of powders with large specific surface areas.

The number of turnovers of a catalyst before it dies is clearly the best definition of the *life* of the catalyst. In practice, this turnover can be very large, say 10^6 or more. As to the turnover frequency, it is frequently of the order of one per second. With much smaller values, the rate is too small to be measured, or to be practical. With much higher values, the rate is too large and becomes influenced by transport phenomena in catalyst pores (see Chapter 6). In fact, reaction temperature is frequently adjusted so as to obtain this commonly encountered value of the activity, $v_t = 1 \ s^{-1}$ (Burwell and Boudart, 1974).

1.26 Selectivity

With more than one reaction taking place, catalyst selectivity is defined, for two reactions, as the ratio of their rates:

$$S = \frac{v_1}{v_2} \tag{1.2.11}$$

Selectivity is frequently easier to measure correctly than individual rates. This is particularly so if the catalyst is readily poisoned in a way in which it is very difficult to control. Besides, selectivity is of intrinsic interest for considerations of mechanisms, as shall be seen later (Maurel et al., 1975).

1.27 Quantities Related to Reaction Rate

In flow reactors, a number of quantities are related to the reaction rate. The first one is defined in terms of the volumetric flow rate and the catalyst volume:

$$\text{space velocity} = \frac{F}{V} \ s^{-1}$$

Frequently this quantity is expressed in h^{-1} and is of the order of $1 \ h^{-1}$ in large-scale processing. The inverse of the *space velocity* is the *space time*:

$$\tau = \frac{V}{F} \ s$$

As to the *space time yield*, it is the quantity of a product produced per quantity of catalyst per unit time. A similar useful quantity can be defined as the *site time yield*: it is the number of molecules of a product made per site in the reactor per second. If the reactor is well-stirred (CSTR, continuous-stirred-tank-reactor), the site time yield is proportional to the turnover frequency as defined above.

1.3 ELEMENTARY STEP, REACTION PATH, AND OVERALL REACTION

1.3 Elementary Step

The stoichiometric equation (1.2.1) does not tell us how the chemical transformation takes place at the molecular level, unless it stands for an elementary step. But if this is indeed the case, the arbitrary choice of a set of stoichiometric coefficients or of a multiple thereof is not permitted, as the elementary step must be written as it takes place at the molecular level.

Consider as an example the dissociative adsorption of dioxygen on two adjacent free sites, each being denoted by the symbol $*$:

$$O_2 + 2* \underset{\overleftarrow{v}}{\overset{\overrightarrow{v}}{\rightleftharpoons}} 2O*$$

By contrast, one may not write:

$$\tfrac{1}{2}O_2 + * \rightleftharpoons O*$$

as this has no mechanistic meaning. The identification of the active site $*$ is the central problem of heterogeneous catalysis.

The net rate of the elementary step is:

$$v = \overrightarrow{v} - \overleftarrow{v}$$

The step may be *reversible* when $\overrightarrow{v} \simeq \overleftarrow{v}$, in which case the equation for the step will be written with a double arrow:

$$\text{Reactants} \rightleftharpoons \text{Products}$$

Or the step may be *irreversible*, if $\overrightarrow{v} \gg \overleftarrow{v}$. Then the equation is written with a single arrow:

$$\text{Reactants} \longrightarrow \text{Products}$$

Finally the step may be *equilibrated*, or quasi-equilibrated, if $\overrightarrow{v} = \overleftarrow{v}$ (or nearly so), in which case a symbol denoting zero net rate (personal communication from Professor K. Tamaru) is used in the equation:

$$\text{Reactants} \rightleftharpoons\!\!\!\ominus\!\!\!\rightleftharpoons \text{Products}$$

For an overall reaction, the equal sign is used:

$$\text{Reactants} = \text{Products}$$

with $\rightleftharpoons$ if the reaction is equilibrated and $\Rightarrow$ if it is far from equilibrium.

1.32 Reaction Path and Stoichiometric Number

A catalytic cycle is defined by a closed sequence of elementary steps. In the first one an active site is converted to an active intermediate. In each subsequent step, the active intermediate is converted into another one. In the last step, the last active intermediate regenerates the free active site.

If we sum up, side by side, the stoichiometric equations for each step, we obtain the stoichiometric equation for the overall equation, provided we take each step σ times, σ being the *stoichiometric number* of that step.

1.33 Single Path Reaction (Temkin, 1971)

As an example, consider the oxidation of SO_2 on a platinum catalyst. The catalytic sequence may consist of two elementary steps: a dissociative adsorption of dioxygen followed by an Eley-Rideal step in which gaseous SO_2 reacts with adsorbed oxygen:

$$
\begin{array}{lcl}
 & & \dfrac{\sigma}{N^{(1)}} \\
O_2 + 2* & \rightleftharpoons 2O* & 1 \\
SO_2 + O* & \rightleftharpoons SO_3 + * & 2 \\
\hline
2SO_2 + O_2 & = \quad 2SO_3 &
\end{array}
$$

If we sum up side by side the equations for both steps, we obtain the stoichiometric equation for the overall reaction if we multiply each step by its stoichiometric number, in this case the couple $(1, 2)$ corresponding to a single path $N^{(1)}$. Another couple $(1/2, 1)$ would correspond to the choice of the stoichiometric equation for the overall reaction:

$$\tfrac{1}{2}O_2 + SO_2 = SO_3$$

Both couples are of course equivalent: there exists only one single path for the reaction. The proper choice of values of σ eliminates the reaction intermediates. The turnover frequency is the number of turnovers per second and per site, along the single reaction path:

$$N = \frac{1}{S}\frac{dn}{dt}(s^{-1})$$

The stoichoimetric numbers were first introduced by Horiuti (1957, 1967).

1.34 Multiple-Path Reaction

There is only one overall reaction, but it may proceed by different paths. An old example is the reaction between H_2 and O_2 on platinum to produce water as first studied by Faraday (1844), and then by a very large number of investigators (Hanson and Boudart, 1978). The next example shows two reaction paths $N^{(1)}$ and $N^{(2)}$ leading to the same overall reaction. The first path $N^{(1)}$ goes through Langmuir-Hinshelwood elementary steps where two adsorbed species react together. The second path $N^{(2)}$ involves an Eley-Rideal step, i.e., a reaction between a surface species and a gaseous species.

	σ	
	$N^{(1)}$	$N^{(2)}$
$O_2 \ + 2* \rightleftharpoons 2O*$	1	1
$H_2 \ + 2* \rightleftharpoons 2H*$	2	0
$H* \ + O* \rightleftharpoons OH* \ + *$	2	0
$OH* + H* \rightleftharpoons H_2O \ + 2*$	2	0
$H_2 \ + O* \rightleftharpoons H_2O \ + *$	0	2
$\overline{2H_2 \ + O_2 \ = \ 2H_2O}$		

1.35 Reaction Networks

In this case, different reaction paths lead to different overall reactions, but the paths $N^{(i)}$ are coupled. As an example, consider the manufacture of water gas, an old process which is attracting attention again for the production of "syngas," CO and H_2:

	σ	
	$N^{(1)}$	$N^{(2)}$
$C* \ + H_2O \rightleftharpoons CO* + H_2$	1	1
$CO* + C \ \rightleftharpoons CO \ + C*$	1	0
$CO* + CO \longleftarrow CO_2 + C*$	0	1
$\overline{C \ + H_2O \ = \ CO \ + H_2}$	$N^{(1)}$	
$CO \ + H_2O \ = \ CO_2 + H_2$		$N^{(2)}$

A rate of reaction is defined along each reaction path. In each step, we recognize stable species like C, i.e., reactants and products and Bodenstein reactive intermediates, like C*. The difference between C and C* from a mechanistic standpoint is that the former is bound to more carbon atoms than the latter (Holstein and Boudart, 1981).

Besides the material balance between reactants and products expressed by the stoichiometric equations, it is necessary to write a material balance for the reaction intermediates.

Thus for the water synthesis reaction, we have a relation between the surface number densities:

$$[L] = [*] + [O*] + [H*] + [OH*]$$

We shall now see how the total specific surface area of a catalyst can be determined (§1.4) or how the number density of metallic atoms can be measured. The aim in both cases is to determine $[L]$, or rather S, the number of active sites, or a number proportional to it.

1.4 Determination of the Total Specific Surface Area of a Catalyst: Method of Brunauer, Emmett and Teller, (BET)

1.41 List of Symbols

V_{ads}	: volume of N_2 (NTP) adsorbed by the sample
V_m	: volume of N_2 (NTP) corresponding to the formation of a monolayer
P	: N_2 pressure in equilibrium with the surface
P_o	: saturated vapor pressure of N_2 at liquid N_2 temperature, 77.3 K with $P_o = 760$ torr
σ	: number density of molecules taken up by the solid, cm^{-2}
σ_m	: number density of adsorbed molecules to form a monolayer, cm^{-2}
θ_i	: fraction of the solid surface covered by a pile of i molecules ($i = 1, 2, \ldots, \infty$)
θ_o	: fraction of the solid surface uncovered by any molecule
V_{coll}	: number of molecules striking the surface, $cm^{-2} s^{-1}$
$\bar{v}$	: mean molecular speed, in vapor
$\bar{v}_x$	: component of mean molecular speed along the x direction
n	: number density of gas molecules, cm^{-3}
n_o	: number density of saturated vapor molecules, cm^{-3}

k : Boltzmann's constant, $1.380662 \times 10^{-23} \, JK^{-1}$

R : gas constant, $8.31441 \, JK^{-1} \, \text{mol}^{-1}$

m : molecular mass

M : mass per mole

$\vec{v}_{ads}$: rate of adsorption

$\overleftarrow{v}_{des}$: rate of desorption

$\alpha, \alpha_2, \ldots, \alpha_\infty$: condensation probability of a molecule striking the surface covered with $1, 2, \ldots, \infty$ layers of molecules

α_o : adsorption probability of a molecule striking the solid surface

k_1 : desorption rate constant for molecules forming a pile one molecule thick

$k, k_2, \ldots, k_\infty$: desorption rate constant for molecules forming a pile, 2 or more molecules thick

G : molecule in the vapor

G_a : adsorbed gas molecule

1.42 Adsorption Isotherm of N_2 at 77.3 K (Normal Boiling Point of Liquid N_2)

One of the most frequent of the five types of adsorption isotherms, V_{ads} versus P/P_o, is shown on Fig. 1.1. The type of isotherm depends on the porosity of the solid and the relative value of the heat of adsorption and the heat of liquefaction of N_2 (Brunauer et al., 1938).

The relative pressure range for the calculation of the total surface area of the solid is known empirically to be between 0.05 and 0.35.

1.43 Some Results From Gas Kinetic Theory (Barrow, 1973)

Gas kinetic theory shows that the mean molecular velocity is:

$$\bar{v} = \left(\frac{8kT}{\pi m}\right)^{\frac{1}{2}} = \left(\frac{8RT}{\pi M}\right)^{\frac{1}{2}} \tag{1.4.1}$$

Another result of the theory is that the number of gas molecules striking a surface per unit area and per unit time is:

$$\bar{v}_{coll} = \frac{1}{4}\bar{v}n \tag{1.4.2}$$

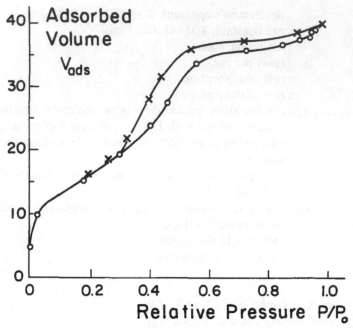

Fig. 1.1 Adsorption of N_2 at 77.3 K on η-alumina
o: increasing pressures
x: decreasing pressures.

This formula applies to the area of an opening through which a molecular beam is formed as well as to collisions with a solid or liquid surface. The diameter of the hole must be small as compared to the mean free path in the gas. Then the mean velocity of the effusing molecules in the x direction is given by:

$$\bar{v}_x = \left(\frac{2kT}{\pi m}\right)^{\frac{1}{2}} = \frac{1}{2}\bar{v} \tag{1.4.3}$$

If instead of the hole, we consider a solid surface, it can be shown that if the number density of molecules is n in the gas, $n/2$ are moving in the x direction, so that $\bar{v}_{coll}$ is given by:

$$\bar{v}_{coll} = \frac{n}{2}\bar{v}_x = \frac{1}{4}\bar{v}n = n\left(\frac{kT}{2\pi m}\right)^{\frac{1}{2}} = n\left(\frac{RT}{2\pi M}\right)^{\frac{1}{2}}$$

A molecule striking the surface may be adsorbed with a probability α. The rate of adsorption $\vec{v}_{ads}$ is then related to the rate of collisions by means of the relation:

$$\vec{v}_{ads} = \alpha \vec{v}_{coll} = \alpha \frac{\bar{v}}{4} n \qquad (1.4.4)$$

1.44 Multilayer Adsorption

With the symbols just defined, the number density of molecules adsorbed in multilayers is:

$$\sigma = \sigma_m \sum_{i=1}^{\infty} i\theta_i \qquad (1.4.5)$$

whereas the surface coverage must be:

$$1 = \theta_o + \sum_{i=1}^{\infty} \theta_i \qquad (1.4.6)$$

The multilayer adsorption isotherm that is sought after is a relation between σ and $(n/n_s) = (P/P_o) = x$. Equations (1.4.5) and (1.4.6) will give the answer provided that θ_o and θ_i ($i = 1, 2, \ldots, \infty$) can be expressed in terms of x. Let us see how that can be done.

1.45 Multilayer Adsorption Isotherm

At equilibrium, each layer is established by a dynamic equilibrium of adsorption and desorption: $G \rightleftarrows G_a$ (see de Boer, 1953).

a) First layer (Fig. 1.2). The rate of adsorption on the uncovered surface as given by gas kinetic theory:

$$\vec{v}_{ads} = \alpha_o \left(\frac{\bar{v}}{4} n \right) \theta_o \quad \text{(adsorbed molecules)} \quad \text{cm}^{-2}\,\text{s}^{-1} \qquad (1.4.7)$$

must be balanced at equilibrium by the rate of desorption of the molecules in the first layer:

$$\vec{v}_{des} = k_1 \sigma_m \theta_1 \qquad (1.4.8)$$

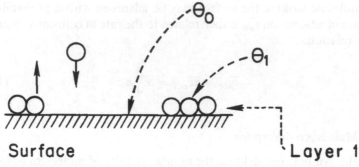

Surface

Fig. 1.2 First layer

Hence, at equilibrium:

$$G \rightleftharpoons G_a$$

$$\alpha_o \frac{\bar{v}}{4} n\theta_o = k_1 \sigma_m \theta_1 \qquad (1.4.9)$$

b) Second layer (Fig. 1.3). Equation (1.4.9) still applies, and the balance for the second layer gives:

$$\alpha_1 \frac{\bar{v}}{4} n\theta_1 = k_2 \sigma_m \theta_2 \qquad (1.4.10)$$

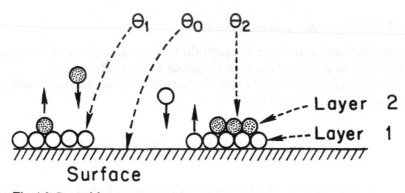

Surface

Fig. 1.3 Second layer

c) Infinite number of layers. The multilayer can be assimilated to a liquid in equilibrium with its saturated vapor. The dynamic equilibrium between rates of condensation and evaporation gives:

$$\alpha_\infty n_s \frac{\bar{v}}{4} \theta_\infty = k_\infty \sigma_m \theta_\infty \qquad (1.4.11)$$

where $\theta_\infty = 1$. If we now assume:

$$\alpha = \alpha_1 = \alpha_2 \cdots = \alpha_\infty \neq \alpha_o \qquad (1.4.12)$$

$$k = k_2 = \cdots = k_\infty \neq k_1 \qquad (1.4.13)$$

to acknowledge the difference in thermodynamic and kinetic properties between a molecule directly adsorbed on the surface and a molecule adsorbed on top of another one, then (1.4.10) and (1.4.11) become:

$$\begin{cases} \alpha \dfrac{v}{4} n\theta_1 = k\sigma_m \theta_2 & (1.4.14) \\[2ex] \alpha \dfrac{\bar{v}}{4} n_s = k\sigma_m & (1.4.15) \end{cases}$$

Division of these last two equations, side by side, gives:

$$\frac{n}{n_s} = \frac{\theta_2}{\theta_1} = \frac{P}{P_o} = x \qquad (1.4.16)$$

Thus we get an iteration formula:

$$\theta_2 = x\theta_1$$

$$\theta_3 = x^2\theta_1$$

which leads to:

$$\boxed{\theta_i = x^{i-1}\theta_1} \qquad (1.4.17)$$

We now need a relation between θ_o and θ_1, which can be obtained by division side by side of (1.4.9) and (1.4.15):

$$\theta_1 = \frac{\alpha_o}{\alpha} \frac{k}{k_1} \frac{n}{n_s} \theta_o$$

Thus:

$$\theta_1 = Cx\theta_o \qquad (1.4.18)$$

Let us write:

$$C = \frac{\alpha_o}{\alpha} \frac{k}{k_1}$$

This leads to the relation between θ_i and θ_o:

$$\boxed{\theta_i = Cx^i\theta_o} \qquad (1.4.19)$$

All that remains to be done to obtain the BET isotherm is to replace θ_i by the value just found in the fundamental relations (1.4.5) and (1.4.6) describing multilayer adsorption:

$$\sigma = \sigma_m \sum_{i=1}^{\infty} i\theta_i$$

$$\sigma = \sigma_m C\theta_o \sum_{i=1}^{\infty} ix^i \qquad (1.4.20)$$

$$1 = \theta_o\left(1 + C \sum_{i=1}^{\infty} x^i\right) \qquad (1.4.21)$$

Hence the adsorbed amount σ depends on x as follows:

$$\sigma = \frac{\sigma_m C \sum_{i=1}^{\infty} ix^i}{1 + C \sum_{i=1}^{\infty} x^i} \qquad (1.4.22)$$

But the denominator contains:

$$\sum_{i=1}^{\infty} x^i = (1 + x^1 + x^2 + \cdots + x^i) - 1$$

which can be evaluated easily since for $x < 1$:

$$\sum_{i=1}^{\infty} x^i \simeq \frac{1}{1 - x} - 1 = \frac{x}{1 - x}$$

As to the numerator, it contains:

$$\sum_{i=1}^{\infty} ix^i = x(1 + 2x + 3x^2 + \cdots + ix^{i-1})$$

the value of which is also found in books on mathematical data. Hence:

$$\sum_{i=1}^{\infty} ix^i = \frac{x}{(1 - x)^2}$$

The expression for σ becomes more simply:

$$\sigma = \frac{\sigma_m Cx/(1 - x)^2}{1 + [Cx/(1 - x)]} \qquad (1.4.23)$$

or alternatively:

$$\sigma = \frac{\sigma_m Cx}{(1 - x)(1 - x + Cx)}$$

This expression can be linearized in its useful form:

$$\boxed{\frac{x}{\sigma(1 - x)} = \frac{1}{C\sigma_m} + x\frac{C - 1}{C\sigma_m}} \qquad (1.4.24)$$

From the slope and intercept of the straight line representing the adsorption data, it is easy to obtain σ_m, the number density of molecules of N_2 forming a monolayer. Many investigations have led to the value of 0.162 nm^2 as the average area per N_2 molecule. Thus the specific area of the solid can be obtained readily.

If we note that $\sigma/\sigma_m = V_{ads}/V_m$ and $x = P/P_o$, the linearized expression of the BET adsorption isotherm becomes:

$$\frac{P}{V_{ads}(P_o - P)} = \frac{1}{CV_m} + \frac{C - 1}{CV_m}\frac{P}{P_o} \qquad (1.4.25)$$

where V_{ads} and V_m are volumes at NTP conditions. The data of Fig. 1.1 are represented on Fig. 1.4 in this manner. The surface area of the sample is obtained by noting that 1 cm^3 (NTP) for V_m corresponds to 4.374 m^2.

In spite of the simplicity of the BET model and of the simplifications made in the derivation of the BET isotherm, the BET specific surface area

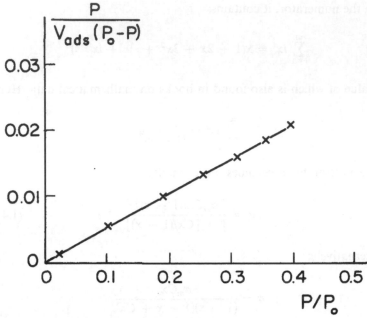

Fig. 1.4 Linearized BET adsorption isotherm

is a correct and reproducible value for most catalytic materials. The BET method is used universally.

1.5 DETERMINATION OF THE NUMBER OF EXPOSED METAL ATOMS

1.51 Principle of the Measurement

Metallic catalysts exist in the form of bulk metal (powder, wire, film) with one or several *promoters*, or in a dispersed state on a nonmetallic porous support (alumina, silica, alumina-silica, zeolites, titania, magnesia, carbon, etc.). Dispersion achieves a high specific surface area of the metal which can be maintained during catalyst activation, use, or regeneration. One of the functions of the support is to prevent growth of the metallic crystallites, a phenomenon improperly called sintering. A typical sample of platinum on alumina (Pt/Al_2O_3 for short) is shown on Fig. 1.5.

Whereas the multilayer physisorption of N_2 or Kr is used to determine the total surface area of the solid, selective chemisorption with formation of a monolayer has been used systematically to determine the surface area of metals. What is measured is the total number of accessible surface atoms, which is not necessarily equal to the number of catalytic sites in

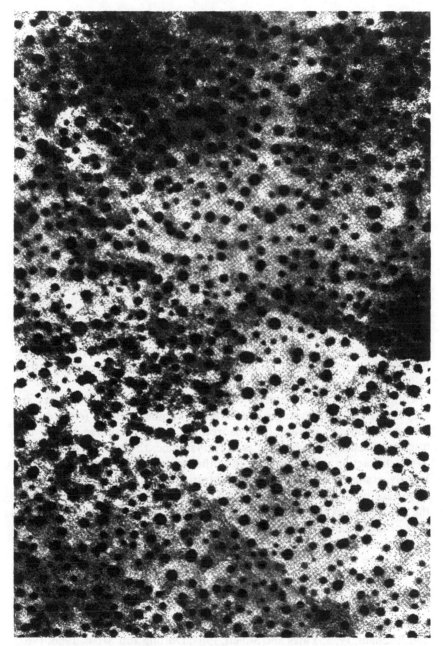

Fig. 1.5 Sample of Pt/Al_2O_3: 1 mm = 2.5 nm: the metallic particles appear as
spots with high contrast. Electron photomicrograph, courtesy of Exxon
Research and Engineering Company

a given reaction, as pointed out before (§1.25). The *dispersion D* of the metal ($D \leq 1$) is defined as the ratio of the number of surface metal atoms to the total number of metal atoms in the sample (Boudart, 1969). Dispersion is also expressed as a percentage. According to IUPAC rules, the expression *"percentage exposed"* should be preferred over the word dispersion.

For a method of selective chemisorption to be successful, the number of surface atoms must be simply equal to the number of molecules chemisorbed on the metal only, multiplied by a factor that expresses the stoichiometry of the chemisorption, which may or may not be dissociative.

The most common adsorptives used in selective chemisorption are H_2, CO, O_2, N_2O, and C_6H_6 (Scholten, 1979). First of all, the correct experimental conditions for the formation of a monolayer must be determined: temperature, pressure, time required for equilibrium. Usually, the measurements must be corrected for, because of adsorption on the support (Benson and Boudart, 1965), absorption inside the metal (Benson et al., 1973), or a weak chemisorption on the metal itself (Sinfelt, 1973). Since percentage exposed, particle size, and surface area are related, it is recommended, whenever possible, to check the results of the selective chemisorption by means of physical techniques: X-ray diffraction line broadening and electron microscopy (Spenadel and Boudart, 1960; Dorling and Moss, 1966, 1967; Cormack and Moss, 1969).

Unfortunately, these physical tools fail when the metallic clusters are very small. Also the averages are different: number average (electron microscopy), surface average (chemisorption), volume average (X-ray diffraction). Besides, if a particle is polycrystalline, the size determined by X-ray diffraction will be smaller than that given by electron microscopy. Finally, assumptions must be made to translate number of atoms exposed to surface or size of the particles (§1.53). From the viewpoint of catalysis, the percentage exposed is the most *direct* and *pertinent* measurement. If part of the metal surface is poisoned, the percentage exposed may correlate directly with catalytic activity (Leclercq and Boudart, 1981), although particle size translated into dispersion would not be a useful quantity.

1.52 Examples of Determination of the Number of Surface Atoms

a) Selective chemisorption of CO (Emmett and Brunauer, 1937; Pliskin and Eischens, 1958; Yates and Sinfelt, 1967; Dorling and Moss, 1966, 1967; Charcosset et al., 1967; Baddour et al., 1968; Cormack and Moss, 1969; Lam and Boudart, 1977). The first use of selective chemisorption is that of Emmett and Brunauer (1937) who used CO to determine the number of iron atoms at the surface of ammonia synthesis catalysts. The stoichiometry in that case was

$$CO + 2Fe_s \longrightarrow \begin{matrix} Fe_s \\ \diagdown \\ \diagup \\ Fe_s \end{matrix} C{=}O$$

where subscript s denotes a surface atom.

In the case of Pt or Pd, a 1:1 stoichiometry is used at not-too-low pressures of CO:

$$CO + Pd_s \longrightarrow Pd_s{=}C{=}O \quad \text{(linear form)}$$

although the stoichiometry corresponding to the bridged form has been used for Pd at low pressure.

b) Selective chemisorption of H_2. Palladium presents a complication as hydrogen dissolves in the bulk with formation of α and β hydrides (Smith, 1948; Boudart and Hwang, 1975). According to Benson et al. (1973), this difficulty can be circumvented by evacuating the sample at room temperature (decomposition of the β phase without desorption) after preliminary adsorption and absorption, or operating at 373 K so that the β phase is not formed. Dihydrogen is widely used to determine the number of surface atoms of Group VIII metals (Spenadel and Boudart, 1960; Dorling and Moss, 1966; Cormack and Moss, 1969). The chemisorption is dissociative, and it is generally admitted that, if the temperature is around 300 K and the equilibrium pressure not too low, a 1:1 surface hydrogen-metal stoichiometry prevails (Sinfelt and Yates, 1968; Wilson and Hall, 1970; Barbaux et al., 1970):

$$H_2 + 2Pt_s \longrightarrow 2Pt_sH$$

The stoichiometry may involve more than one H per Pt_s when the particle size becomes very small (<1.5 nm) but the situation is not always clear. There is an abundant literature on this controversial subject, old (e.g., Benton, 1926; Sieverts and Bruning, 1931) and new (Sinfelt and Yates, 1968; Mears and Hansford, 1967; Wilson and Hall, 1970).

c) Selective chemisorption of O_2. The chemisorption is dissociative on most metals, and a frequently postulated stoichiometry is:

$$O_2 + 2Pt_s \longrightarrow 2Pt_sO$$

But this is not universally agreed upon. Again there are discrepancies in the case of very small particles ($O:Pt_s = \frac{1}{2}:1$) according to Wilson and Hall, 1970; Dalla Betta and Boudart, 1973. In the presence of poisons which may also chemisorb oxygen, e.g., sulfur on platinum, oxygen chemisorption may lead to incorrect values of the percentage of platinum exposed (Leclercq and Boudart, 1981).

d) Titration by O_2 and H_2 (Gruber, 1962; Benson and Boudart, 1965; Barbaux et al., 1970; Vannice et al., 1970). This method is frequently used in the case of fresh platinum catalysts. The most common version is the titration of preadsorbed oxygen by H_2:

$$Pt_s + \tfrac{1}{2}O_2 = Pt_sO$$

$$Pt_sO + \tfrac{3}{2}H_2 = Pt_sH + H_2O$$

As seen above, the 1:1 stoichiometries Pt_s:H and Pt_s:O may fail at very small particle sizes. But there may be a compensation as more hydrogen and less oxygen seems to be chemisorbed on very small particles. At any rate, the method is convenient, as no preliminary reduction of the metal at high temperature is necessary, as in the case of hydrogen chemisorption alone. In spite of uncertain stoichiometries, the method seems to yield values of the dispersion of platinum in excellent agreement with those obtained by other methods. Its disadvantage over chemisorption of H_2 is that it adds the uncertainty of O_2 chemisorption. Its advantage is sensitivity. Indeed, in volumetric determination (Benson and Boudart, 1965), three atoms of hydrogen are picked up per surface platinum atom versus only one in ordinary chemisorption of H_2. In a gravimetric determination (Barbaux et al., 1970), the weight change between that of the oxygen-covered sample and the hydrogen-covered sample is much easier to measure than that between an evacuated sample and one covered with hydrogen. But there are problems if the surface of platinum is poisoned, as mentioned above in the case of oxygen chemisorption alone (Leclercq and Boudart, 1981).

In spite of the remaining uncertainties, the dispersion of platinum and many other supported metals can be measured in a reproducible manner in different laboratories. The conditions of sample preparation and measurement must be strictly adhered to. Only ultra-pure gases should be used, especially in flow methods, and the assumed stoichiometry must be spelled out.

e) Selective chemisorption of N_2O. In the case of copper, oxygen is too strongly held and forms mutilayers readily if one starts from dioxygen. By contrast, hydrogen and carbon monoxide are held too weakly. A good compromise is N_2O, which decomposes as follows:

$$N_2O + 2Cu_s \rightleftharpoons (Cu_s)_2O + N_2$$

This takes place between 363 and 373 K (Osinga et al., 1967; Scholten and Konvalinka, 1969; Echevin and Teichner, 1975).

1.53 How to Obtain the Surface Area of the Metal

Provided that the stoichiometry of the selective chemisorption is agreed upon, the number of accessible surface atoms can be obtained as illustrated above, and from that number it is easy to obtain the dispersion or percentage exposed.

To obtain the surface area of the metal, it is necessary to calculate [L], the number density of surface atoms. What is frequently done is to assume that low index planes are equally represented. Thus for platinum, the value of [L] is taken to be the arithmetic mean between the values of the number density of the (100), (110) and (111) planes. This is not unreasonable, but it is arbitrary. Typical values for $[L]/10^{15}$ cm^{-2} are 1.10, 1.15, and 1.7 for platinum, iridium, and copper respectively.

The specific surface area of the metal is not a very useful quantity anyway. If it merits being reported at all, it is only as a stepping stone toward the estimation of a particle size (see §1.54). The surface area per gram of metal should be reported and not the surface area per gram of catalyst.

1.54 How to Obtain the Size of the Metal Particle

This is a useful quantity, not for direct use in catalysis, but for comparison between values of selective chemisorption with values obtained by physical methods, X-ray line broadening, X-ray scattering, or electron microscopy.

The specific surface area of the metal S_M is first obtained from the selective chemisorption data by assuming an average surface number density. Then the average particle size d is obtained by means of:

$$d = f \frac{V_{sp}}{S_M}$$

where V_{sp} is the specific volume of the metal and f is a shape factor which is generally unknown. In fact, much remains to do to determine the shape of small particles of metals on supports, the change of shape with the nature of the support, or on a given support with the adsorption of gases or reacting molecules. The values of f collected in Table 1.1 are between 1 and 6. In the absence of any other information, the value $f = 6$ is not unreasonable.

From the viewpoint of catalysis, all that matters is the dispersion D, or the percentage exposed, without any further interpretation. In fact, the turnover frequency is very easily obtained by dividing specific rate (say in μmol per gram of catalyst per second) by the specific uptake of gas

TABLE 1.1   Relation between shape and size

Shape of the particle	Dimension	Factor f
Cube	Edge	5 or 6
Sphere	Diameter	6
Cylinder	Diameter	4
Plate	Thickness	1 or 2

used in the selective chemisorption, adjusted for dissociation and surface stoichiometry (in μmol per gram of catalyst).

Nevertheless, if a comparison with results from physical techniques is wanted, or just out of habit of dealing with particle size, a handy conversion between dispersion D (expressed as a ratio) and particle size d in nm is given by:

$$D \simeq 0.9/d \quad (d \text{ in } nm)$$

or more simply

$$\boxed{D = \frac{1}{d/nm}}$$

This is a good approximation for the transition metals with particles of spherical shape (Boudart and Hwang, 1975). For further details on supported metal catalysts, the reader should consult the book of Anderson (1975).

1.6 DEFINITION OF CATALYSIS

1.61 Decomposition of Ozone in the Gas Phase

In the presence of oxygen atoms, ozone decomposes through the elementary step:

$$O + O_3 = 2O_2$$

The rate of this uncatalyzed reaction (in $cm^{-3} s^{-1}$) is simply in terms of the number density of O and O_3:

$$v_{nc} = k[O][O_3] \tag{1.6.1}$$

with an experimentally known rate constant:

$$k = 1.9 \times 10^{-11} \exp(-2300/T) \, cm^3 \, s^{-1} \tag{1.6.2}$$

The quantity 2300, expressed in Kelvin, is a *temperature of activation*, i.e., the activation energy divided by the gas constant. If every hard sphere collision were reactive, the rate would be given by the expression from the kinetic theory of gases:

$$v = \pi \bar{v} \sigma^2 [O][O_3] \tag{1.6.3}$$

where σ is a mean collision cross-section:

$$\sigma = (\sigma_O + \sigma_{O_3})/2$$

and:

$$\bar{v} = \left(\frac{8kT}{\pi\mu}\right)^{\frac{1}{2}}$$

with μ, the reduced mass:

$$\mu = m_O m_{O_3}/(m_O + m_{O_3})$$

Comparison between (1.6.1) and (1.6.3) shows an order of magnitude agreement between:

$$\pi \bar{v} \sigma^2 \simeq 10^{-10} \, cm^3 \, s^{-1}$$

and the pre-exponential factor of k. Thus if the decomposition of ozone at, say 200 K is quite slow, it is because the exponential factor containing the activation energy.

1.62 Decomposition of Ozone Catalyzed by Atomic Chlorine

By definition of catalysis (§1.32) we must have a closed sequence with Cl consumed in the first step and regenerated in the last:

$$\begin{array}{l}
Cl \; + O_3 \xrightarrow{k_1} O_2 + ClO \\
\underline{ClO + O \xrightarrow{k_2} O_2 + Cl} \\
O \; + O_3 \; = \; 2O_2
\end{array}$$

with rate constants, as determined experimentally:

$$k_1 = 5 \times 10^{-11} \exp(-140/T) \, \text{cm}^3 \, \text{s}^{-1} \qquad (1.6.4)$$

$$k_2 = 1.1 \times 10^{-10} \exp(-220/T) \, \text{cm}^3 \, \text{s}^{-1} \qquad (1.6.5)$$

The concentration of the active centers is a constant:

$$[L] = [\text{Cl}] + [\text{ClO}] \qquad (1.6.6)$$

At the steady state, as will be seen later (§1.64), the rate of the catalyzed reaction v_c is given by:

$$v_c = \frac{k_1 k_2 [L][O][O_3]}{k_1 [O_3] + k_2 [O]} \qquad (1.6.7)$$

In the present case, $k_1 \simeq k_2$ but $[O] \ll [O_3]$. Hence:

$$v_c = k_2 [L][O] \qquad (1.6.8)$$

Let us now compare the rates of the catalyzed and uncatalyzed reactions:

$$\frac{v_c}{v_{nc}} = \frac{k_2 [L]}{k[O_3]}$$

but if we assume:

$$[L]/[O_3] = 10^{-3}$$

we get:

$$\frac{v_c}{v_{nc}} = \frac{k_2}{k} \times 10^{-3}$$

Hence:

$$\frac{v_c}{v_{nc}} = 5.79 \times 10^{-3} \exp(2080/T) \qquad (1.6.9)$$

At $T = 200$ K, v_c/v_{nc} is equal to 190. This is the enhancement of the rate as a result of the presence of the catalyst.

The original catalyst Cl reacts with the first reactant to give an *intermediate* ClO which regenerates Cl by reacting with the second reactant. The energy profile (Fig. 1.6) of the catalytic cycle is quite instructive. Curve *a* corresponds to the uncatalyzed reaction: an appreciable activation barrier must be surmounted. If the system follows curve *c*, the activation barrier to form the intermediate is very shallow, but the intermediate is too stable, so that its further reaction will necessitate going over a sizable energy of activation. Profile *b* corresponds to that of a good catalyst: the intermediate is stable enough so that it can be formed readily, but not too stable so that it can be decomposed easily. This is the golden rule of catalysis following Sabatier. The catalysis of the ozone decomposition by chlorine atoms, which takes place in the stratosphere, illustrates Sabatier's principle. Note that a single atom, namely chlorine, is a catalyst. No molecular species, cluster or solid, is necessary.

An energy diagram of the form of Fig. 1.6 should be available for a quantitative understanding of a catalytic cycle.

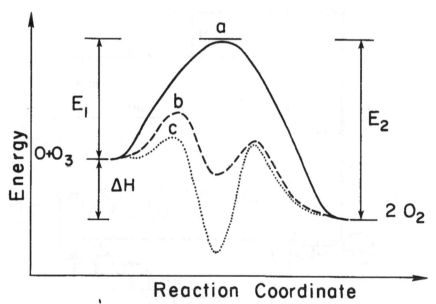

Fig. 1.6 Potential energy versus reaction coordinate
a: uncatalyzed reaction
b: case of a good catalyst
c: case of a bad catalyst
ΔH: enthalpy of reaction equal to $E_2 - E_1$

1.63 Energy Profile for a Catalytic Reaction

It is frequently said or written that, in spite of more than sixty years of research, we do not "understand" the mechanism of ammonia synthesis on commercial iron catalysts. What is meant by "understanding"? Perhaps a good measure of understanding can be provided by a quantitative thermochemical kinetic profile of the kind discussed in the case of the catalytic decomposition of ozone (Fig. 1.6).

The merit of such a knowledge in the case of chain reactions has been emphasized by Benson (1977). For catalytic reactions, theorists are currently trying to calculate many of the missing values of energy involved in the steps of the catalytic cycle (Rappé and Goddard, 1981).

For the case of ammonia synthesis on Fe (111), Ertl (1981) has proposed a thermochemical kinetic profile, as shown on Fig. 1.7. There are still

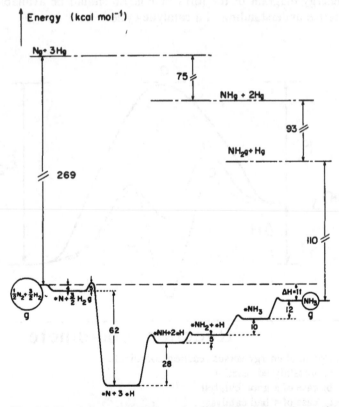

Fig. 1.7 Energy versus reaction coordinate for ammonia synthesis
X_g: gaseous species
$X*$: adsorbed species

several uncertainties in this profile. Maybe it can be said that ammonia synthesis will be well understood when the last major uncertainties in this profile will have disappeared.

1.64 Quasi-Steady State Approximation

The way to obtain a rate expression for a catalytic reaction is to apply the Bodenstein theory of the steady state approximation to the elementary steps in the catalytic cycle (Boudart, 1968; Christiansen, 1953; Giddings and Shin, 1961).

a) General treatment. Consider a sequence of two irreversible first-order steps with rate constants k_1 and k_2.

$$A \xrightarrow{k_1} B \xrightarrow{k_2} C$$

Note that other reactants and products may be involved besides A, B and C, but they do not appear in the rate equations.

Denote by $[A]$, $[B]$, $[C]$ the concentrations of A, B, and C at any time and $[A]_o$ the concentration of A at time $t = 0$, at which $[B]_o = [C]_o = 0$. Then define:

$$\frac{[A]}{[A]_o} = x \qquad \frac{[B]}{[A]_o} = y \qquad \frac{[C]}{[A]_o} = z$$

which are related simply:

$$x + y + z = 1 \tag{1.6.10}$$

The differential equations describing the evolution of the system with time are:

$$\frac{dx}{dt} = -k_1 x \tag{1.6.11}$$

$$\frac{dy}{dt} = k_1 x - k_2 y \tag{1.6.12}$$

$$\frac{dz}{dt} = k_2 y \tag{1.6.13}$$

It follows that:

$$\frac{dx}{dt} + \frac{dy}{dt} + \frac{dz}{dt} = 0 \tag{1.6.14}$$

The solution of (1.6.11), with $x = 1$ at $t = 0$ is:

$$x = \exp(-k_1 t) \qquad (1.6.15)$$

Substitution of (1.6.15) into (1.6.12) gives a first-order linear differential equation:

$$\frac{dy}{dt} + k_2 y = k_1 \exp(-k_1 t) \qquad (1.6.16)$$

whose solution is straightforward:

$$y = \frac{k_1}{k_2 - k_1} \left[\exp(-k_1 t) - \exp(-k_2 t) \right] \qquad (1.6.17)$$

Integration of (1.6.13) gives the value of z:

$$z = 1 - \frac{k_2}{k_2 - k_1} \exp(-k_1 t) + \frac{k_1}{k_2 - k_1} \exp(-k_2 t) \qquad (1.6.18)$$

The concentration of A decreases exponentially while that of B rises, goes through a maximum, then decreases (Fig. 1.8). At the maximum:

$$\frac{dy}{dt} = 0$$

Hence:

$$t_{max} = \frac{1}{k_2 - k_1} \ln \frac{k_2}{k_1}$$

and:

$$y_{max} = \left[\frac{k_1}{k_2} \right]^{k_2/(k_2 - k_1)} \qquad (1.6.19)$$

For $k_1 = 2k_2$, $t_{max} = (\ln 2)/k_2$ and $y_{max} = 0.5$ whereas $x = z = 0.25$.

Figure 1.8 shows the evolution in time of the concentrations. Note that an induction time is required before B reaches its maximum value, at $t =$

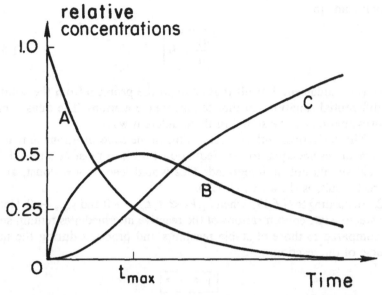

Fig. 1.8 Concentrations as a function of time for two consecutive first-order reactions with $k_1 = 2k_2$. Note that the slope of the C curve at time zero is equal to zero. If this were not the case, there would be an indication of a bypass linking A and C directly, or perhaps of diffusional limitations (see Chapter 6).

t_{max}. At that time, the C curve goes through an inflection point at which $d^2z/dt^2 = 0$.

Suppose now that B is a very reactive intermediate, i.e., $k_2 \gg k_1$. At times that are not too short, (1.6.7) can then be simplified to:

$$y = \frac{k_1}{k_2} \exp(-k_1 t) \cong \frac{k_1}{k_2} x \qquad (1.6.20)$$

Hence:

$$k_2 y - k_1 x \cong 0$$

Or, because of (1.6.11) and (1.6.13):

$$\frac{dz}{dt} + \frac{dx}{dt} \cong 0 \qquad (1.6.21)$$

which leads to:

$$\boxed{\frac{dy}{dt} \cong 0}$$ (1.6.22)

This is a fundamental result that changes the problem from the solution of differential equations to that of algebraic equations. The quasi-steady approximation can be stated in three different ways:

1. The derivatives with respect to time of the concentrations of reactive intermediates are equal to zero (equation 1.6.22). It must be stressed that (1.6.22) should not be integrated. This would lead to $y = \text{const}$, an incorrect result, as shown by (1.6.20).

2. According to (1.6.19), when $k_1/k_2 \ll 1$, $t_{max} \to 0$ and $y_{max} \to 0$. Hence, the steady-state concentrations of the reactive intermediates remain small as compared to those of stable reactants and products during the main course of the reaction:

$$\boxed{y \ll x, z}$$

Also, the induction time is usually quite small as compared to the reaction time.

3. Finally, equation (1.6.21) means that in a reaction sequence at the steady state, the rates of all steps involving Bodenstein intermediates are equal:

$$\boxed{\frac{dx}{dt} = -\frac{dz}{dt}}$$

b) Application to the catalytic decomposition of ozone. Just stating that the rates of the two steps in the catalytic sequence section 1.62 are equal at the steady state, we get:

$$k_1[Cl][O_3] = k_2[ClO][O]$$

which together with the conservation equation:

$$[L] = [Cl] + [ClO]$$

leads to the steady-state concentration of the intermediate:

$$[ClO] = \frac{k_1[L][O_3]}{k_1[O_3] + k_2[O]}$$

Hence:

$$v = k_2[\text{ClO}][\text{O}]$$

$$v = \frac{[L]k_1k_2[\text{O}][\text{O}_3]}{k_1[\text{O}_3] + k_2[\text{O}]}$$

which is the result mentioned earlier.

In this example the steady-state approximation is justified by the second condition mentioned above, as the concentration of chlorine atoms has been assumed to be much smaller than that of ozone.

It will be seen later how the steady-state approximation is used to treat catalytic cycles of any complexity. In any situation, the approximation is valid only after an induction period. Intuitively, it would appear that this induction time should be of the order of magnitude of the *turnover time*, i.e., the inverse of the turnover frequency. This result has indeed been obtained by Temkin (1976) for a two-step catalytic cycle. At any rate, a reaction cannot be considered as catalytic until the catalyst has turned over more than once. Otherwise the reaction must be considered as stoichiometric. It was mentioned in section 1.25 that frequently the turnover frequency is of the order of $1 \ s^{-1}$. In that case, the induction time should be quite short. But if the turnover frequency were $10^{-3} \ s^{-1}$ for an admittedly slow catalytic reaction, the induction period could be as long as twenty minutes, and before that time has elapsed after the start of the reaction, it would be risky to talk about a "catalyst" or a "catalytic reaction." In fact, the use of the word catalyst to denote any solid sample of "promise" is to be deplored. The word should be reserved for a proven catalyst, i.e., one that has turned over a few times. How many times is, like beauty, in the eye of the beholder.

REFERENCES

Anderson, J. R. 1975. *Structure of Metallic Catalysts.* New York: Academic Press.

Baddour, R. F., Modell, M., and Heusser, U. K. 1968. *J. Phys. Chem.* 72:3621.

Barbaux, Y., Roger, B., Beaufils, J. P., and Germain, J. 1970. *J. Chim. Phys.* 67:1035.

Barrow, G. M. 1973. *Physical Chemistry.* New York: McGraw-Hill. 3rd ed. (1976), p. 41.

Bénard, J. 1970. *Catal. Rev.* 3:93.

Benson, J. E. and Boudart, M. 1965. *J. Catal.* 4:704.

Benson, J. E., Hwang, H. S., and Boudart, M. 1973. *J. Catal.* 30:146.

Benson, S. W. 1977. *Thermochemical Kinetics.* New York: John Wiley & Sons.

Benton, A. F. 1926. *J. Am. Chem. Soc.* 48:1850.

Boer, J. H. de 1953. *The Dynamical Character of Adsorption.* Oxford: Oxford University Press.

Boudart, M. 1968. *Kinetics of Chemical Processes.* Englewood Cliffs, N.J.: Prentice-Hall, Inc.

Boudart, M. 1969. *Advan. Catal. Relat. Subj.* 20:153.

Boudart, M. 1972. *AIChE J.* 18:465.

Boudart M., Collins, D. M., Hanson, F. V., and Spicer, W. E. 1977. *J. Vac. Sci. Technol.* 14:441.

Boudart, M. and Hwang, H. S. 1975. *J. Catal.* 39:44.

Brunauer, S., Emmett, P. H., and Teller, E. 1938. *J. Am. Chem. Soc.* 60:309.

Burwell, R. L., Jr., and Boudart, M. 1974. In *Investigations of Rates and Mechanisms of Reactions*, Part 1, chap. 12, ed. E. S. Lewis. New York: John Wiley & Sons.

Charcosset, H., Barthomeuf, D., Nicolova, R., Revillon, A., Tournayan, L., and Trambouze, Y. 1967. *Bull. Soc. Chim. Fr.*, 4555.

Christiansen, J. A. 1953. *Advan. Catal. Relat. Subj.* 5:311.

Cormack, D. and Moss, R. L. 1969. *J. Catal.* 13:1.

Dalla Betta, R. and Boudart, M. 1973. *Proc. 5th Intl. Cong. Catalysis*, Palm Beach, 1972, ed. Hightower, p. 1329. North Holland: Elsevier.

Djéga-Mariadassou, G., Marques, A. R., and Davignon, L. 1982. *J. Chem. Soc. Faraday I* 78:2447.

Dorling, T. A. and Moss, R. L. 1966. *J. Catal.* 5:111.

Dorling, T. A. and Moss, R. L. 1967. *J. Catal.* 7:378.

Echevin, B. and Teichner, S. J. 1975. *Bull. Soc. Chim. Fr.*, 1487.

Emmett, P. H. and Brunauer, S. 1937. *J. Am. Chem. Soc.* 59:310 and 1553.

Ertl, G. 1981. *Proc. 7th Intl. Cong. Catalysis*, ed. T. Seiyama and K. Tanabe, Part A, p. 21. Tokyo: Kodansha.

Faraday, M. 1844. *Experimental Researches in Electricity.* London.

Giddings, J. C. and Shin, H. K. 1961, *Trans. Faraday Soc.* 57:468.

Glasstone, S., Laidler, K. J., and Eyring, H. 1941. *The Theory of Rate Processes.* New York: McGraw-Hill.

Gruber, H. L. 1962. *J. Phys. Chem.* 66:48.

Hanson, F. V. and Boudart, M. 1978. *J. Catal.* 53:56.

Holstein, W. L. and Boudart, M. 1981. *J. Catal.* 72:328.

Horiuti, J. 1957. *J. Res. Inst. Catalysis*, Hokkaido Univ. 5:1.

Horiuti, J. and Nakamura, T. 1967. *Advan. Catal. Relat. Subj.* 17:1.

IUPAC. 1976. Definitions, terminology and symbols in colloid and surface chemistry, Part 2. *Heterogeneous Catalysis, Pure Appl. Chem.* 45:71.

IUPAC. 1979. Manual of symbols and terminology for physicochemical quantities and units. *Pure Appl. Chem.* 51:1.

Lam, Y. L. and Boudart, M. 1977. *J. Catal.* 50:530.

Langmuir, I. 1916. *J. Am. Chem. Soc.* 38:2221.

Leclercq, G. and Boudart, M. 1981. *J. Catal.* 71:127.

Maurel, R., Leclercq, G., and Barbier, J. 1975. *J. Catal.* 37:324.

Mears, D. E. and Hansford, R. C. 1967. *J. Catal.* 9:125.

Osinga, T. J., Linsen, B. G., and Van Beek, W. P. 1967. *J. Catal.* 7:277.

Pliskin, W. A. and Eischens, R. P. 1958. *Advan. Catal. Relat. Subj.* 10:1.

Ponec, V. and Sachtler, W. M. H. 1972. *Proc. 5th Intl. Cong. Catalysis*, p. 645.

Rappé, A. K. and Goddard, W. A. 1981. In *Potential Energy Surfaces and Dynamics Calculations*, ed. D. J. Truhlar, p. 661. New York: Plenum.

Scholten, J. J. F. 1979. In *Preparation of Catalysts, II*, ed. B. Delmon, P. Grange, P. Jacobs, and G. Poncelet, p. 685. Amsterdam: Elsevier.

Scholten, J. J. F. and Konvalinka, J. A. 1969. *Trans. Faraday Soc.* 65:2465.

Sieverts, A. and Bruning, H. Z. 1931. *Anorg. Chem.* 201:136.

Sinfelt, J. H. 1973. *J. Catal.* 29:308.

Sinfelt, J. H., Lam, Y. L., Cusumano, J. A., and Barnett, A. E. 1976. *J. Catal.* 42:227.

Sinfelt, J. H. and Yates, D. J. C. 1968. *J. Catal.* 10:362.

Smith, D. P. 1948. *Hydrogen in Metals.* Chicago: Chicago University Press.

Spenadel, L. and Boudart, M. 1960. *J. Phys. Chem.* 64:204.

Sundquist, B. E. 1964. *Acta Metallurgica* 12:67.

Taylor, H. S. 1925. *Proc. Roy. Soc.* (London) A 108, 105.

Temkin, M. I. 1957. *Zhur. Fiz. Khim.* 31:1.

Temkin, M. I. 1965. *Dok. Akad. Nauk. SSSR* 161, 160.

Temkin, M. I. 1971. *Int. Chem. Eng.* 11:709.

Temkin, M. I. 1976. *Kin. Kat.* 17:1095.

Temkin, M. I. 1979. *Advan. Catal. Relat. Subj.* 28:173.

Vannice, M. A., Benson, J. E., and Boudart, M. 1970. *J. Catal.* 16:348.

Wagner, C. 1970. *Advan. Catal. Relat. Subj.* 21:323.

Wilson, G. R. and Hall, W. K. 1970. *J. Catal.* 17:190.

Yates, D. J. C. and Sinfelt, J. H. 1967. *J. Catal.* 8:348.

Chapter 2

KINETICS OF ELEMENTARY STEPS: ADSORPTION, DESORPTION, AND SURFACE REACTION

2.1 INTRODUCTION

A major development in heterogeneous catalysis is the very recent accumulation of data on the kinetics of elementary steps on well-defined surfaces. These are atomically clean, smooth planes on large single crystals. Cleanliness is achieved by high temperature evacuation, treatment with reactive gases, and/or ion bombardment. The structure of the surface is established by low-energy electron diffraction. Its composition is checked and monitored by Auger electron spectroscopy. All observations are made with the sample in an ultrahigh vacuum chamber with a background pressure of the order of 10^{-8} to 10^{-10} Pa. Residual gas composition is monitored by mass spectroscopy. Very pure gases are leaked into the sample, at pressures of 10^{-4} to 10^{-7} Pa. Although studies have been conducted on various substances, we shall confine our attention mostly, but not exclusively, to metals which are emphasized throughout this monograph.

A popular model of single crystal surfaces (Fig. 2.1) evolved from investigations on crystal growth goes back to early work of Stranski (1928, 1931) and Kossel (1957), elaborated by Hirth and Pound (1963).

A single crystal surface exhibits *terraces T* which are low Miller index planes with high surface density separated by *ledges L* which may have *kinks K*. A terrace between two ledges is a step. Defects on terraces consist of *vacancies* or adsorbed atoms (*adatoms*). Thus the smooth surface may indeed be rough on an atomic scale. This is the model called *TLK*. The systematic use of stepped single crystals with regularly spaced ledges has been popularized by Somorjai (1981). These surfaces are effectively planes of high Miller index.

For very small metallic particles, called clusters, planes do not have a meaning. It is better to define surface structure with the notations introduced by Van Hardeveld and Hartog (1969). They consider the coordination number i of a surface *atom* and the coordination number of a

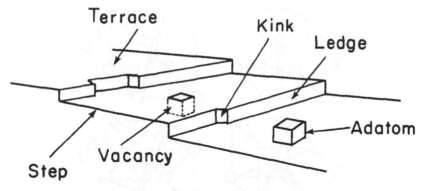

Fig. 2.1 *Schematic representation of a single crystal surface*

surface *site*. A surface atom is denoted by C_i when it has i nearest neighbors. A site is denoted by B_j when it has j nearest neighbors. Examples of C_4, C_6, and C_7 atoms are shown on Fig. 2.2. Several sites, B_4, B_5, B_6, and B_7 are represented on Fig. 2.3.

Many catalytic studies have been carried out on metallic wires, gauze or low specific surface area powders, besides supported metals. A comparison between these data and those obtained on a large single crystals has now become possible. Studies on the latter have also been conducted at high

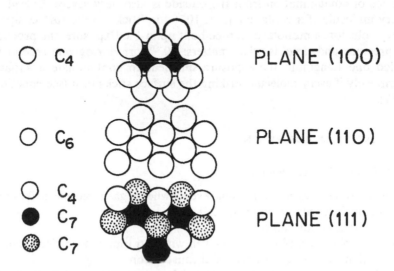

Fig. 2.2 Surface atoms C_i for low Miller index planes of a body-centered cubic lattice.

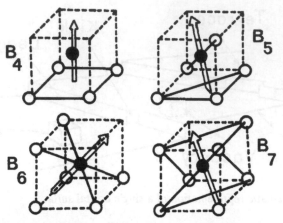

Fig. 2.3 Surface sites B_j on planes of a body-centered cubic lattice.

pressures. Large single crystals are ideal model systems, and data obtained with them are now recognized as standards.

One of the reasons for such a comparison between the various experimental data is that surface contamination is always to be contended with in catalysis by solids. Impurities can accumulate at the surface from within or from the outside. The chance of contamination from within is the higher, the lower the dispersion of the metal; with single crystals, the chance of contamination from the outside is also very severe. Indeed, if every molecule of a contaminant at 10^{-6} torr sticks to the surface upon every collision, a monolayer can be built up in 1 s. Exposure, the product of pressure and time, is what matters: 10^{-6} torr during one second is called one Langmuir, the exposure sufficient to contaminate a surface completely if every molecule striking the surface sticks to it (see equation 1.4.2).

2.2 KINETICS OF ADSORPTION

2.21 Lennard-Jones Diagram

Consider the succession of possible events when a gas molecule strikes the surface of a stepped crystal (Fig. 2.4).

Two main events can occur:

1. Trapping by a physisorption well with a trapping coefficient α, into a state that may be a *precursor* of chemisorption.

2. Dissociative or non-dissociative adsorption into a chemisorption well after eventual diffusion of the molecule along the surface.

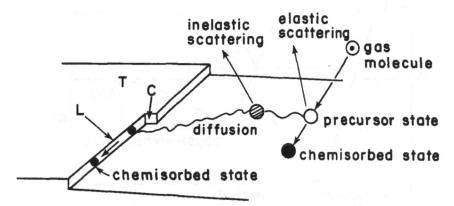

Fig. 2.4 Adsorption of a gas molecule
⊙ : gas molecule
○ : physisorbed molecule
● : chemisorbed molecule

Adsorption is frequently depicted by a diagram of Lennard-Jones (1932) representing the potential energy of the system as a function of the distance between the molecule and the surface or, more properly, the distance along a reaction coordinate on a multidimensional surface (Fig. 2.5). The physisorbed and chemisorbed states appear in a shallow and deep well respectively. Access to the chemisorption well may necessitate surmounting an activation barrier E_a. The latter may be equal to zero, as its value depends on the relative values of ΔH_p and ΔH_c as well as the relative position of the potential energy curves, as fixed by r_p and r_c.

Fig. 2.6 shows two types of energy profiles for adsorption. The first one (Fig. 2.6a) corresponds to the physisorption of methane on the (100) face of tungsten with a rather deep physisorption well of 29 kJ mol^{-1}, which could be a precursor state for activated chemisorption at higher temperatures (Yates and Madey, 1971). Fig. 2.6b illustrates the dissociative chemisorption of H_2 on Cu (Balooch et al., 1974). Similarly, the chemisorption of N_2 on the (100) and (110) faces of iron is also clearly dissociative and activated (Bozso et al., 1977).

The concept of activated chemisorption has been controversial ever since it was introduced by Taylor (1931), who was struck by the fact that chemisorption is frequently a slow phenomenon. Tompkins (1979) states that for simple molecules on clean metals, activated adsorption does not exist ($E_a = 0$).

But there exist now clear examples of activated adsorption for simple molecules such as H_2 and N_2, on copper ($E_a = 12 \ldots 20$ kJ mol^{-1},

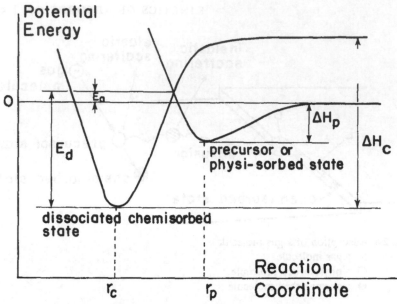

Fig. 2.5 Lennard-Jones diagram for dissociative chemisorption

E_a: activation barrier for transition between physisorption and chemisorption

E_d: activation energy for desorption

ΔH_p and ΔH_c: enthalphy of physisorption and chemisorption respectively

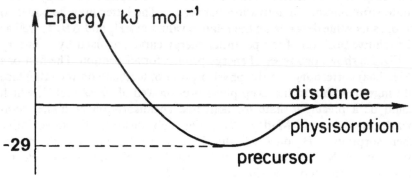

Fig. 2.6a Physisorption of methane on W (100), following data of Yates and Madey (1971)

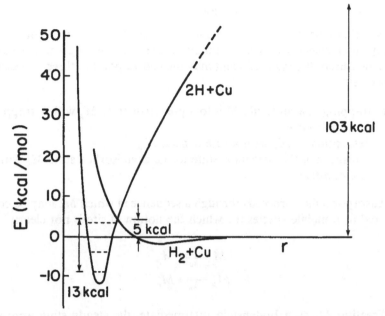

Fig. 2.6b Chemisorption of hydrogen on copper following data of Balooch et al. (1974)

Balooch et al., 1974) and the (110) face of iron ($E_a = 29 \; kJ \; mol^{-1}$, Bozso et al., 1977). The old controversy has been laid to rest.

It is convenient to express adsorption rate by means of a sticking coefficient, s, defined at a given value of surface coverage, as the net rate of adsorption v_a of a molecule M divided by the rate of collision of M with the surface.

The trapping coefficient α for physisorption or adsorption into a precursor state should not be confused with the sticking coefficient s, which may describe a sequence of events (Fig. 2.4).

For a bare surface (subscript o):

$$s_o = \frac{v_{a,o}}{v_c} = \frac{v_{a,o}}{\frac{\bar{v}}{4}[M]}$$

(2.2.1)

Measured values of s_o range from 10^{-6} to unity. Consider now the values of s on bare and covered surfaces, with the help of the kinetic theory of gases and transition state theory.

2.22 Adsorption on a Bare Surface

a) Sticking coefficient: kinetic theory of gases. Adsorption is not necessarily an elementary process but may proceed through a precursor state. The measured sticking coefficient may then correspond to three successive processes:

1. trapping of a molecule M into a precursor state M_p with a trapping rate constant k_c.
2. desorption of M_p with a rate constant k_d.
3. transition of the precursor state to the chemisorbed state M_a with a rate constant k_a.

Adsorption thus proceeds through a sequence in which M_p may be considered as a mobile species for which the notion of site is not clear:

$$M \underset{k_d}{\overset{k_c}{\rightleftarrows}} M_p$$

$$M_p \xrightarrow{\ k_a\ } M_a$$

Treating M_p as a Bodenstein intermediate, the steady state approximation yields:

$$k_c[M] - k_d[M_p] = k_a[M_p]$$

Hence:

$$[M_p] = \frac{k_c[M]}{k_a + k_d}$$

and the rate of adsorption $v_{a,o}$ is:

$$v_{a,o} = \frac{k_a k_c[M]}{k_a + k_d} \tag{2.2.2}$$

With s_o and α being respectively the sticking and trapping coefficients, on a clean surface, the rate constant k_c is obtained from the kinetic theory of gases:

$$k_c = \alpha \frac{\bar{v}}{4} \tag{2.2.3}$$

and by definition:

$$s_o = \frac{k_a k_c}{k_a + k_d} \times \frac{4}{\bar{v}}$$
(2.2.4)

Replacing k_c by its value (2.2.3):

$$\boxed{s_o = \frac{\alpha}{1 + (k_d/k_a)}}$$
(2.2.5)

The dependence of s_o on temperature depends on that of α. The latter is related to the behavior of the thermal accommodation coefficient, a.c., of inert gases (Weinberg and Merrill, 1971). If a gas at temperature T_g^o strikes a surface (e.g., a filament) at temperature T_s, heat transfer will take place. The gas leaves the surface at a temperature T_g which is equal to T_s for complete thermal accommodation. Hence, a thermal accommodation coefficient can be defined:

$$a.c. = \lim_{T_g^o \to T_s} \left[\frac{T_g - T_g^o}{T_s - T_g^o} \right]$$

Weinberg and Merrill have pointed out the relations between α and a.c. It is found experimentally that when T_s goes to zero, s_o goes to α which behaves like an a.c. Hence the trapping coefficient seems to be determined by energy transfer between gas and surface.

When the adsorbed molecules are simple and the chemisorption is non-activated, the sticking coefficient on bare surfaces varies between 10^{-3} and unity on transition metals. One may ask whether these values correspond to those of conventional catalytic reactions for which the rate of turnover is $1 \ s^{-1}$. Suppose that, in the reaction, adsorption takes place at ambient conditions. Then $[M] \cong 10^{19} \ cm^{-3}$ and:

$$v_c = \frac{\bar{v}}{4}[M] \cong 10^4 \times 10^{19} \ cm^{-2} \ s^{-1}$$

For the customary total density of sites, in order of magnitude, $[L] = 10^{15} \ cm^{-2}$, we get

$$\frac{v_c}{[L]} = 10^0 \ s^{-1}$$

and

$$s_o = \frac{N}{v_c/[L]} = 10^{-8} \tag{2.2.6}$$

a value which is many orders of magnitude smaller than those mentioned on the preceding page. We must conclude that in normal catalysis the surface is not bare, or there exists a substantial activation barrier to adsorption. Both eventualities probably happen together simultaneously in catalytic reactions.

b) Sticking coefficient: transition state theory. It is interesting, for further developments, to check some of the results above by means of transition state theory. The rate of collisions of M with a surface is given by:

$$v_c = \frac{kT}{h} \exp(\Delta S_M^{o\dagger}/R) \exp(-\Delta H_M^{o\dagger}/RT) [M] \tag{2.2.7}$$

where all quantities have their usual meaning, with $\Delta S_M^{o\dagger}$ and $\Delta H_M^{o\dagger}$ being the standard entropy and enthalpy of formation of the transition state from the gas molecule M. Standard states are 1 molecule per cm^3 in the gas and 1 molecule per cm^2 at the surface. To evaluate $\Delta S_M^{o\dagger}$, we reason that reaching the transition state from a hard-sphere molecule M corresponds simply to the loss of one translational degree of freedom: tr, 3D → tr, 2D:

$$\Delta S_M^{o\dagger} = S_{M,tr,2D}^{o\dagger} - S_{M,tr,3D}^{o\dagger} \tag{2.2.8}$$

This can be assessed readily (e.g., Boudart, 1975):

$$\Delta S_M^{o\dagger}/R = \left(2 + 2 \ln \frac{kT/h}{\bar{v}/4} \right) - \left(\frac{5}{2} + 3 \ln \frac{kT/h}{\bar{v}/4} \right)$$

thence:

$$\boxed{\exp(\Delta S_M^{o\dagger}/R) = \frac{\bar{v}}{4} \frac{h}{kT} \exp(-\tfrac{1}{2})} \tag{2.2.9}$$

Next, to evaluate $\Delta H_M^{o\dagger}$, we first note that $\Delta H_M^{o\dagger} = \Delta U_M^{o\dagger}$ since the collision neither destroys nor creates a new molecular species. Hence:

$$\Delta H_M^{o\dagger} = \Delta U_M^{o\dagger} = \Delta U_o^{o\dagger} + \int_0^T \Delta c_p^{\dagger} dT$$

But for a collision, the variation of internal energy at 0 K, $\Delta U_M^{o\ddagger}$ is equal to zero. Thus (see Barrow, 1973)

$$\boxed{\Delta H_M^{o\ddagger} = -\tfrac{1}{2}RT}$$ (2.2.10)

Substitution into (2.2.7) yields:

$$v_c = \left\{\frac{kT}{h}\left(\frac{\bar{v}}{4}\frac{h}{kT}\exp - \tfrac{1}{2}\right)\exp\frac{\tfrac{1}{2}RT}{RT}\right\}[M]$$

or finally:

$$v_c = \frac{\bar{v}}{4}[M]$$

the classical result from transition state theory. We shall remember that it corresponds to the loss of one translational degree of freedom.

c) Rate of adsorption $v_{a,o}$. According to transition state theory, on a bare surface (fraction of surface coverage, $\theta = 0$) with a number density of free sites $[L]$, the rate of adsorption in one elementary step:

$$* + M \longrightarrow M *$$

is given by

$$v_{a,o} = \frac{kT}{h}\left\{\exp[(S_M^{o\ddagger} - S_M^o)/R]\exp(-\Delta H_M^{o\ddagger}/RT)\right\}[M][L] \quad (2.2.11)$$

We will now assume that $M*$ is localized on a site and that the transition state to reach it is immobile, i.e., has no translational degree of freedom left. Since M is not a hard-sphere molecule, we must then separate the *internal* vibrational and rotational contributions to the entropy from the external translational ones. Hence:

$$S_M^{o\ddagger} - S_M^o = (S_{M,\text{int}}^\ddagger - S_{M,\text{int}}) - S_{M,\text{tr},3D}^o - S* \quad (2.2.12)$$

where $S*$ is the entropy of vibration of the site. Let us define:

$$\Delta S_{\text{int}}^\ddagger = S_{M,\text{int}}^\ddagger - S_{M,\text{int}} - S* \quad (2.2.13)$$

To compare results from transition state theory with those of gas kinetic theory, let us note that:

$$\exp[(S^o_{M,tr,2D} - S^o_{M,tr,3D})/R] = \frac{h}{kT}\frac{\bar{v}}{4}\exp(-\tfrac{1}{2}) \qquad (2.2.14)$$

Thus the entropy factor in (2.2.11) can be written as:

$$\exp[(S^{o\dagger}_M - S^o_M)/R] = \frac{h}{kT}\frac{\bar{v}}{4}\exp[-\tfrac{1}{2}]\exp[(-S^o_{M,tr,2D}/R)]\exp(\Delta S^{o\dagger}_{int}/R)$$

$$(2.2.15)$$

Let us now evaluate $\Delta H^{o\dagger}_M$:

$$\Delta H^{o\dagger}_M = \Delta U^{o\dagger}_o + \int_0^T \Delta c^\dagger_p \, dT = \Delta U^{o\dagger}_o - \tfrac{1}{2}RT \qquad (2.2.16)$$

With all substitutions made in equation (2.2.11), we get:

$$v_{a,o} = [L]\frac{\bar{v}}{4}\exp(-S^o_{M,tr,2D}/R)\exp(\Delta S^{o\dagger}_{int}/R)\exp(-E/RT)[M] \qquad (2.2.17)$$

where we have replaced $\Delta U^{o\dagger}_o$ by the Arrhenius activation energy E. On the other hand, according to (2.2.1):

$$v_{a,o} = s_o \frac{\bar{v}}{4}[M]$$

We finally obtain the sought-after expression for the sticking coefficient:

$$\boxed{s_o = P\exp(-E/RT)} \qquad (2.2.18)$$

with:

$$P = \underbrace{\{[L]\exp(-S^o_{M,tr,2D}/R)\}}_{\delta}\underbrace{\{\exp(\Delta S^\dagger_{int}/R)\}}_{B} \qquad (2.2.19)$$

For CO_2 at 1273 K, $\bar{v} = 7.8 \times 10^4$ cm s^{-1}, $kT/h = 2.65 \times 10^{13}$ s^{-1}. With $[L] = 10^{15}$ cm^{-2}, δ is of the order of 10^{-4}. As to B, it must be smaller than or equal to unity, as some entropy is usually lost in the formation of the transition state for adsorption. In conclusion, P must be almost certainly smaller than unity.

TABLE 2.1   Sticking coefficient parameters for CO_2 on metals at 1273 K (Grabke, 1967)

Metal	P	E $(kcal\ mol^{-1})$
Au	3.0	49
Ag	8.5	52
Pd	1.0×10^4	63
Cu	2.4×10^4	61
Ni	1.8×10^4	61

Consider the data of Grabke (1967), in Table 2.1, for adsorption of CO_2 on metals at 1273 K. The quantity P exceeds unity and in three instances by several orders of magnitude. It is therefore likely that the measured process requires more than one elementary step. Values of P larger than unity can be attributed to diffusion (see p. 69) surface impurities (Kemball, 1953), or to a variation of $[L]$ with temperature (Vol'kenshtein, 1949).

Keeping in mind that P and E may have a complex meaning, we note that their values in Table 2.1 constitute our first example of a *compensation effect* introduced by Schwab and Cremer (1955). This effect is such that if E increases, so will the pre-exponential factor, so that the product of the latter and of the exponential factor does not vary too much.

For a rate constant $k = A \exp(-E/RT)$, one may observe a linear relation as shown on Fig. 2.7, where the relation is of the type:

$$\ln A = \text{Const}_1 + \text{Const}_2 \times E \qquad (2.2.20)$$

It is possible to express formally Const_2 as

$$\text{Const}_2 = 1/RT_\theta \qquad (2.2.21)$$

when T_θ has the dimension of a thermodynamic temperature. The rate constant then becomes:

$$k = k_o \exp[(-E/R)(1/T - 1/T_\theta)] \qquad (2.2.22)$$

where $k_o = \exp(\text{Const}_1)$.

Then, for either a reaction on various catalysts, or several reactions on a given catalyst, we can obtain on an Arrhenius plot (Fig. 2.8), a fan of straight lines which intersect at a temperature T_θ called the *isokinetic temperature* since all rate constants have the same value at that temperature.

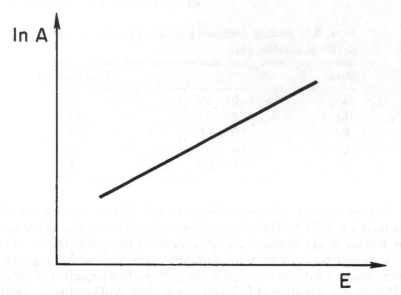

Fig. 2.7 Compensation law (for example, Anderson and Kemball, 1954; Kemball, 1953; Boudart 1961)

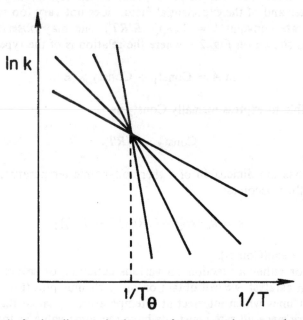

Fig. 2.8 Arrhenius diagram in the case of compensation, equation (2.2.22)

Above T_θ, the more rapid reactions are those with higher values of E, but the converse is true below T_θ. If E and A vary but little, compensation may be ascribed to experimental error. Otherwise, it probably always follows from one or several phenomena. Other examples will be described.

In summary, from (2.2.1), (2.2.18), and (2.2.19) we have:

$$v_{a,o} = \delta B \frac{\bar{v}}{4} \exp(-E/RT)[M]$$

Since δ is of the order of 10^{-3} or 10^{-4} and B is smaller than or equal to unity, the pre-exponential factor of the adsorption rate constant is always smaller than 10^4 cm s^{-1} if the transition state is immobile. It may be equal to 10^4 cm s^{-1} if the transition state preserves translational freedom parallel to the surface.

2.23 Adsorption on a Surface with Coverage θ Different from Zero

What happens to the sticking coefficient s for an arbitrary value of coverage θ?

a) Langmuir's site exclusion principle. According to Langmuir, a molecule striking an occupied site is reflected away from the surface. It is now clear that this model is not universal, especially in the case of clean metal surfaces at low coverage where the striking molecule may be held to the surface in a precursor state long enough to diffuse to a site where chemisorption can take place (Ehrlich 1961, 1963).

If some sites are occupied, $[L]$ in the equations above must be replaced by the concentration of free sites $[*]$, or in terms of the fraction of occupied sites θ:

$$[*] = [L](1 - \theta) \qquad (2.2.23)$$

Thus in the case of non-dissociative adsorption:

$$v_a = v_{a,o}(1 - \theta) \qquad (2.2.24)$$

But in the case of dissociative adsorption,

$$M_2 + ** \Longrightarrow M**M$$

v_a depends on the probability of finding two adjacent sites. It is necessary to replace $[L]$ by $[**]$ in the equation (2.2.11) for the rate of adsorption. Consider the statistical equilibrium between the five possible configurations of free sites or of sites occupied by M (Fig. 2.3), in the absence of any interaction forces:

$$2M** \rightleftharpoons M**M + **$$ (2.2.25)

where $M**$, $M**M$, and $**$ represent respectively a partially occupied, completely occupied, or empty pair of sites. The equilibrium constant (Hill, 1960) is:

$$\frac{[M**M][**]}{[M**]^2} = \frac{1}{4}$$ (2.2.26)

With θ as the fraction of occupied sites, $[L]$ the density of sites, z the number of nearest neighbors equal to 4 for the example of Fig. 2.9, we obtain:

$$\theta_{MM} = \frac{[M**M]}{\frac{1}{2}z[L]}$$

$$\theta_{MO} = \frac{[M**]}{\frac{1}{2}z[L]}$$ (2.2.27)

$$\theta_{OO} = \frac{[**]}{\frac{1}{2}z[L]}$$

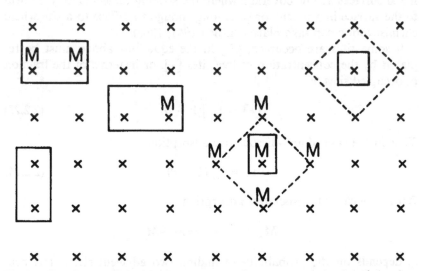

Fig. 2.9 The five possible situations involving free or occupied pairs of sites in a square array

The factor 1/2 in the denominator reflects the indistinguishability of neighboring sites which should not be counted twice. Subscripts O and M refer to sites that are free or occupied by M. As to single sites, we have as usual:

$$\theta \equiv \theta_M = \frac{[M*]}{[L]} \quad \text{and} \quad \theta_o = \frac{[*]}{[L]} \tag{2.2.28}$$

with $\theta_M + \theta_O = 1$.

For these last two types of sites, we have:

$$[M*] = \frac{1}{z}(2[M**M] + 1[M**])$$

$$[*] = \frac{1}{z}(2[**] + 1[M**]) \tag{2.2.29}$$

Hence:

$$\frac{1}{4}\theta_{MO}^2 = \theta_{MM}\theta_{OO}$$
$$\theta_M = \theta_{MM} + \frac{1}{2}\theta_{MO}$$
$$\theta_O = \theta_{OO} + \frac{1}{2}\theta_{MO} \tag{2.2.30}$$

and:

$$\theta_{OO} = \theta_O^2$$
$$\theta_{MM} = \theta_M^2 \tag{2.2.31}$$

It follows readily (Boudart, 1975):

$$[**] = \frac{1}{2}z[L](1 - \theta_M)^2 \tag{2.2.32}$$

and the rate of adsorption for $\theta \neq 0$ is:

$$\boxed{v_a = v_{a,o}(1 - \theta)^2 \frac{z}{2}} \tag{2.2.33}$$

Note that v_a is still proportional to $[L]$, through $v_{a,o}$. It is not proportional to $[L]^2$ as it might appear prima facie.

b) The sticking coefficient in the case of a precursor state. We have already considered the sequence:

$$M \underset{k_d}{\overset{k_c}{\rightleftharpoons}} M_p$$

$$M_p \xrightarrow{k_a f(\theta)} M_a$$

The difference is now that the adsorption rate constant k_a must be multiplied by a certain function $f(\theta)$ of θ, representing the probability to find groups of free adjacent sites. For example, $f(\theta)$ might be $(1 - \theta)^n$. Thus, taking (2.2.5) into account, we get:

$$s = \frac{\alpha}{1 + \dfrac{k_d}{k_a f(\theta)}} = \frac{s_o\left(1 + \dfrac{k_d}{k_a}\right)}{1 + \dfrac{k_d}{k_a f(\theta)}} \tag{2.2.34}$$

A precursor state is more probable at low temperature. Indeed, if $s_o \ll 1$, which corresponds to $k_d/k_a \gg 1$, $s = s_o f(\theta)$, which corresponds to Langmuir's expressions (2.2.24) or (2.2.33) for the rate of adsorption. This case

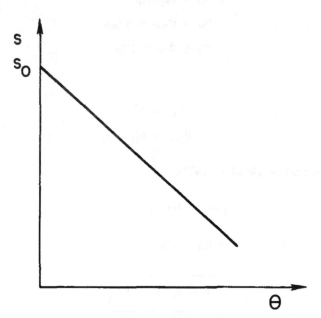

Example 1: Adsorption of N_2 on W (111). The behavior is Langmuirian: $s = s_o (1 - \theta)$. At about 300 K, s_o for N_2 is as high as 0.74 on W (320) but about 10^{-2} on W (110).

corresponds to high temperatures and the existence of an activation barrier to adsorption (high value of E_a). Then the precursor state plays no kinetic role. This is probably the case in most catalytic applications.

The opposite case is $k_d/k_a \ll 1$. In that situation, as observed frequently on sparsely covered surfaces at low temperature, s stays equal to s_0 until relatively high values of θ are reached. This is because adsorption is the last in a series of steps, the first one being the formation of a precursor state. The latter would be important in catalysis only in the eventuality of rapid chemisorption (high k_a, low k_d). Examples 1–4 (King, 1979) illustrate various possibilities.

These examples have been chosen to illustrate the complexity of adsorption kinetics on well-defined clean metal surfaces. But it is clear that the sticking probability always goes down at sufficiently large value of coverage and is small when the adsorption is activated. Hence under usual conditions of heterogeneous catalysis, adsorption kinetics following Langmuir is probably adequate and will be used in what follows.

As to desorption kinetics, a wealth of data has been obtained by means of thermal desorption spectroscopy or temperature programmed desorption. The principles of the method will be described first.

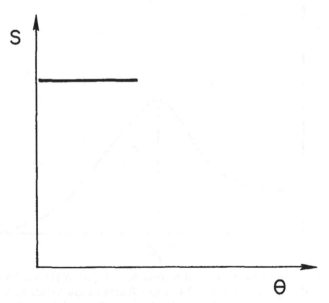

Example 2: Adsorption of N_2 on W (100)
It is found experimentally that s remains approximately constant (~ 0.6 at ~ 300 K) at low values of θ. This is the case $k_d/k_a \ll 1$.

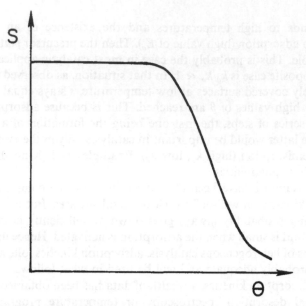

Example 3: Adsorption of N_2 on W (310). This is similar to the previous case. At higher values of θ, s falls rapidly with coverage.

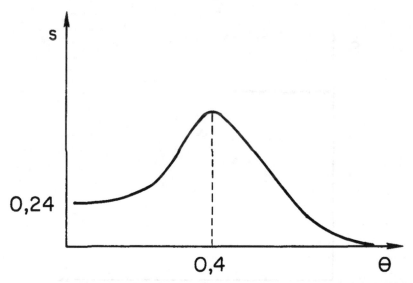

Example 4: In this particular case of adsorption of N_2 on W (110) at 85 K, s goes through a maximum (0.4) before falling to low values at saturation. A cooperative mechanism has been invoked, with formation of *islands* of adsorbate due to attractive interaction.

2.3 KINETICS OF DESORPTION

The temperature programmed desorption is one of the essential experimental kinetic developments since Langmuir. It is an extension of Becker's flash filament technique by Ehrlich (1961, 1963) and Redhead (1962). Developed with single crystals, it has been extended by Cvetanovic and Amenomiya (1967, 1972) to the study of powders. In flash desorption, the heating rate is quite high, whereas in temperature programmed desorption, heating rates can be variable and even quite low. With powders, desorbed species are flushed out by a carrier gas. In what follows, the version of the technique will be presented, as used with single crystal samples in ultrahigh vacuum chambers.

2.31 Experimental Description of the Phenomenon

The phenomenon deals with the desorption of species that have been irreversibly adsorbed at the temperature of adsorption.

Consider a chamber of volume V (cm^3), with a leak valve for feeding gas at a molecular rate F (s^{-1}) and a pump operating at a rate P (cm^3 s^{-1}). A crystal with area A (cm^2) is heated at a linear rate

$$T = T_o + \beta t \qquad (2.3.1)$$

where β is typically of the order of 10 K s^{-1} and t is the time (s). The initial temperature T_o is often between 100 and 300 K. At any time, the number density of molecules in the chamber is n (cm^{-3}) (Fig. 2.10).

After the surface has been cleaned, a steady-state is reached at which the density of molecules in the chamber is n_s with a steady-state density at the surface a_{T_o}. The material balance is:

$$F = n_s P \qquad (2.3.2)$$

As the temperature is raised, desorption starts to occur. It is assumed that no readsorption takes place, either on the crystal or on the walls of the chamber. The new balance is:

$$F + Av_d = nP + V\frac{dn}{dt} \qquad (2.3.3)$$

where v_d is the rate of desorption (cm^{-2} s^{-1}) and $V(dn/dt)$ represents the number of molecules per second causing an increase in pressure in the chamber. Substituting F by its value (2.3.2), denoting $n - n_s$ by $n*$ and

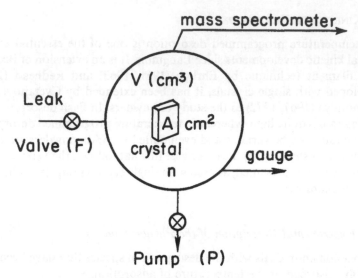

n: molecules cm^{-3}

Fig. 2.10 Apparatus for temperature programmed desorption

noting that $dn/dt = dn*/dt$, we get

$$\frac{A}{V} v_d = n* \frac{P}{V} + \frac{dn*}{dt} \qquad (2.3.4)$$

in which the dimension of A/V is the inverse of a length, $1/L$ and P/V is the inverse of a time τ which is the mean residence time of molecules in the chamber. Thus:

$$\frac{n*}{\tau} + \frac{dn*}{dt} = \frac{v_d}{L} \qquad (2.3.5)$$

If the pumping speed is large enough, τ is small enough and (2.3.5) simplifies to:

$$\boxed{v_d = \frac{L}{\tau} n*} \qquad (2.3.6)$$

where the pressure in the chamber is proportional to $n*$. In this way, as temperature rises, one or several desorption peaks are observed in the desorption spectrum, depending on the binding energy of the molecules with

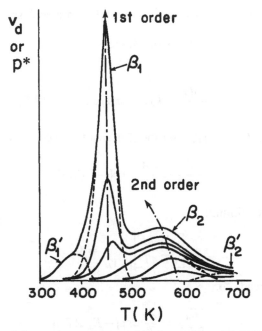

Fig. 2.11 Thermal desorption spectrum of hydrogen from W (100), $p^* = p - p_s$, (Tamm and Schmidt, 1969)

the surface. The thermal desorption spectrum of Fig. 2.11 exhibits four peaks of desorption of hydrogen from the (100) face of a tungsten crystal (Tamm and Schmidt, 1969).

The information contained in these spectra includes: the number of peaks related to the number of binding states, the value of the temperature T_M at the peak maximum, and the number of molecules in each binding state, proportional to the area under each peak.

Further, the analysis of the spectra and their change with the heating rate β or the initial coverage θ_o gives in principle the additional information for each binding state: E_d, the activation energy for desorption; A_d, the pre-exponential factor of the desorption rate constant, k_d; and the kinetic order of the desorption process.

2.32 First-Order Desorption

It corresponds to

$$M* \longrightarrow * + M$$

If σ is the surface density (cm^{-2}) of adsorbed molecules:

$$k_d = A_d \exp(-E_d/RT) \tag{2.3.7}$$

and

$$v_d = -\frac{d\sigma}{dt} = k_d\sigma = \sigma A_d \exp(-E_d/RT) \tag{2.3.8}$$

Noting that $1/dt = \beta/dT$, the rate of desorption becomes:

$$-\frac{d\sigma}{dT} = \frac{\sigma A_d}{\beta} \exp(-E_d/RT) \tag{2.3.9}$$

At the peak maximum:

$$\left[\frac{d}{dT}\frac{d\sigma}{dT}\right]_{T=T_M} = 0 \tag{2.3.10}$$

which easily leads to:

$$\frac{E_d}{RT_M^2} = \frac{A_d}{\beta} \exp(-E_d/RT_M) \tag{2.3.11}$$

or

$$\boxed{2 \ln T_M - \ln \beta = \frac{E_d}{RT_M} + \ln \frac{E_d}{RA_d}} \tag{2.3.12}$$

We see immediately that the value of T_M at a constant value of β is independent of the initial coverage θ_o, a typical feature of first-order desorption kinetics. Further, it can be shown that $\sigma_M = \sigma_o/e$. Finally the straight line $2 \ln T_M - \ln \beta = f(1/T_M)$ obtained by changing β and following the corresponding change in T_M has a slope E_d/R and an intercept which then yields A_d (Madix, 1979a).

2.33 Second-Order Desorption

It corresponds to:

$$2M* \longrightarrow M_2 + 2*$$

Following Redhead (1962), we have, in the absence of readsorption:

$$v_d = -\frac{d\sigma}{dt} = k_d\sigma^2 \tag{2.3.13}$$

or:

$$-\frac{d\sigma}{dT} = \frac{\sigma^2}{\beta} A_d \exp(-E_d/RT) \qquad (2.3.14)$$

At the peak maximum:

$$\left[\frac{d}{dT}\frac{d\sigma}{dT}\right]_{T=T_M} = 0$$

Hence:

$$\frac{E_d}{RT_M^2} = \frac{2(k_d)_M \sigma_M}{\beta} \qquad (2.3.15)$$

It can be shown (Redhead, 1962) that $\sigma_M \simeq \sigma_o/2$. This gives:

$$\boxed{2 \ln T_M - \ln \beta = \frac{E_d}{RT_M} + \ln \frac{E_d}{A_d R \sigma_o}} \qquad (2.3.16)$$

The lines $2 \ln T_M - \ln \beta = f(1/T_M)$ give the same information as before, either by changing β with constant σ_o, or by changing σ_o at constant β. Note, however, that now T_M depends on initial coverage σ_o, proportional to the area of desorption peaks. This verifies that desorption is indeed a second-order process.

Let us also note that Cvetanovic and Amenomiya (1967, 1972) have shown that the desorption peak is asymmetric for a first-order desorption but symmetric for second-order desorption.

We will now show that for second- order desorption:

$$\sigma_M \simeq \sigma_o/2.$$

From (2.3.14), we know that:

$$-\frac{d\sigma}{dT} = \frac{A_d}{\beta} \sigma^2 \exp(-E_d/RT)$$

which gives after integration and regrouping of terms:

$$\int_{\sigma_o}^{\sigma_M} -\frac{d\sigma}{\sigma^2} = \int_{T_o}^{T_M} \frac{A_d}{\beta} \exp(-E_d/RT)\,dT \qquad (2.3.17)$$

Hence:

$$\frac{1}{\sigma_M} - \frac{1}{\sigma_o} = \left[\frac{A_d R}{\beta E_d} T_M^2 \exp(-E_d/RT_M)\right] \times f \qquad (2.3.18)$$

with:

$$f = 1 - \frac{T_o^2}{T_M^2} \exp\left[\frac{-E_d}{R}\left(\frac{1}{T_o} - \frac{1}{T_M}\right)\right] \simeq 1 \qquad (2.3.19)$$

Thus:

$$\frac{1}{\sigma_M} - \frac{1}{\sigma_o} \simeq \frac{A_d R}{\beta E_d} T_M^2 \exp(-E_d/RT_M) \qquad (2.3.20)$$

Using the result of equation (2.3.15), we obtain:

$$\frac{1}{\sigma_M} - \frac{1}{\sigma_o} \simeq \frac{1}{2\sigma_M}$$

or:

$$\sigma_o \simeq 2\sigma_M.$$

Many complications can occur in the interpretation of thermal desorption spectra (Petermann, 1972). Among them are the possibility of readsorption, transition of one binding state to another, or the serious problem of interaction between adsorbed species.

Although experimental refinements continue to appear (Bell and Hecker, 1981), peaks are not always well resolved, and computer analysis of the spectra is mandatory.

In many cases, for simple molecules pre-adsorbed on clean metallic surfaces, measured values of E_d are identical with values of binding energy as the adsorption process is frequently non-activated. In this way, temperature programmed desorption has contributed heavily to the accumulation of adsorption thermochemical data.

Let us now treat the kinetics of unimolecular and bimolecular desorption with the emphasis on pre-exponential factors.

2.4 UNIMOLECULAR SURFACE REACTIONS

As we have seen, for adsorption:

$$v_{a,o} = s_o \frac{\bar{v}}{4}[M] = A \exp(-E_a/RT)[M] \qquad (2:2.1)$$

with $A < 10^4$ cm s^{-1}. For the elementary process:

$$M* \longrightarrow * + M \qquad (2.4.1)$$

transition state theory gives for the rate of desorption v_d:

$$v_d = \frac{kT}{h} \exp(\Delta S^{o\dagger}/R) \exp(-\Delta H^{o\dagger}/RT)[M*] \qquad (2.4.2)$$

with

$$A_d = \frac{kT}{h} \exp(\Delta S^{o\dagger}/R)$$

If there is very little change in molecular structure between $M*$ and $M^\dagger$, the standard entropy of activation would be nearly zero and A_d should be about 10^{13} s^{-1}. Yet values of A_d larger or smaller than 10^{13} s^{-1} are often found. This suggests that $\Delta S^{o\dagger}$ is often smaller or larger than zero. In particular, many values of A_d considerably higher than 10^{13} s^{-1} suggest a gain in rotational freedom as the adsorbed species reaches the transition state (Madix, 1979b).

There is another way to express desorption rates as pertaining to evaporation. As was seen in §1.45, the rate of condensation can be expressed as:

$$\alpha \frac{\bar{v}}{4}[M] \quad \text{(eq. 1.4.11)}$$

with $0 < \alpha < 1$. Since at equilibrium, this is also equal to the rate of vaporization, the maximum value of the latter, for $\alpha = 1$, is given by:

$$\frac{\bar{v}}{4}[M]_e$$

where $[M]_e$ corresponds to the vapor in equilibrium with the condensed phase. This was verified for the vaporization rate of solid and liquid mercury by Volmer and Estermann (1921):

$$v_{\text{vap,max}} = \frac{\bar{v}}{4}[\text{Hg}]_e \qquad (2.4.3)$$

For mercury, α is thus equal to unity. But this is not always the case. Frequently the role of vaporization of a solid is smaller than its maximum value, which may indicate a sequence of elementary steps on a TLK type of stepped crystal (Hirth and Pound, 1963). To handle such cases, a vaporization coefficient α_{vap} can be introduced, similar to the sticking coefficient in adsorption. It is defined by:

$$\alpha_{vap} = \frac{v_{vap}}{v_{vap,max}} = \frac{(\bar{v}/4)[Hg]}{(\bar{v}/4)[Hg]_e} = p/p_e \tag{2.4.4}$$

where p is often called the Langmuir pressure, inferior to the equilibrium pressure p_e.

At equilibrium between liquid and vapor, say for mercury:

$$K = \frac{p_e}{(Hg)_L} = p_e \tag{2.4.5}$$

where $(Hg)_L$ is the activity of pure liquid mercury, equal to unity. Thus p_e can be written as:

$$R \ln p_e = -\frac{\Delta H^o_{vap}}{T} + \Delta S^o_{vap} \tag{2.4.6}$$

p_e being expressed in atmosphere, and ΔH^o_{vap} and ΔS^o_{vap} being respectively the standard enthalpy and entropy of vaporization, the standard state being 1 atm.

Away from equilibrium, by analogy with (2.4.6):

$$R \ln p = -\frac{\Delta H^*_{vap}}{T} + \Delta S^*_{vap} \tag{2.4.7}$$

where ΔH^*_{vap} and ΔS^*_{vap} are quantities that may be considered as enthalpy and entropy of activation, as will be seen shortly. When these quantities are equal to their thermodynamic counterparts ΔH^o_{vap} and ΔS^o_{vap}, the coefficient α_{vap} is equal to unity. Consider a few cases where $\alpha_{vap} \neq 1$.

Example 1: $\Delta H^*_{vap} = \Delta H^o_{vap}$, with $\Delta S^*_{vap} < \Delta S^o_{vap}$ (Mortensen and Eyring, 1960) (eqs. 2.4.4 and 2.4.6):

$$\alpha_{vap} = \exp(\Delta S^*_{vap} - \Delta S^o_{vap})/R \tag{2.4.8}$$

and α_{vap} can be interpreted as a sticking coefficient with a zero activation energy (eq. 2.2.18).

Example 2: $\Delta S^*_{vap} = \Delta S^o_{vap}$, but $\Delta H^*_{vap} > \Delta H^o_{vap}$ (Rosenblatt and Lee, 1968). In this case:

$$\alpha_{vap} = \exp\left(\frac{\Delta H^o_{vap} - \Delta H^*_{vap}}{RT}\right) \qquad (2.4.9)$$

which can be interpreted as an activated desorption. This is the case for the vaporization of arsenic from the (111) face of a single crystal (Rosenblatt and Lee, 1968). In that system, $\alpha_{vap} = 4.6 \times 10^{-5}$ at 550 K ($p_e = 1.44 \times 10^{-5}$ atm; $\Delta H^o_{vap} = 138.4$ kJ mol^{-1}; $\Delta S^o_{vap} = 158.8$ J mol^{-1} K^{-1}; $\Delta H^*_{vap} = 183.5$ kJ mol^{-1}; $\Delta S^*_{vap} = 158.4$ J mol^{-1} K^{-1}). The process is represented schematically on the potential energy profile of Fig. 2.12.

Example 3: $\Delta H^*_{vap} < \Delta H^o_{vap}$ and $\Delta S^*_{vap} < \Delta S^o_{vap}$. This is another example of a compensation effect:

$$p = A \exp(-\Delta H/RT) \qquad (2.4.10)$$

where both pre-exponential factor and ΔH decrease simultaneously as equilibrium conditions are replaced by experimental kinetic ones. This

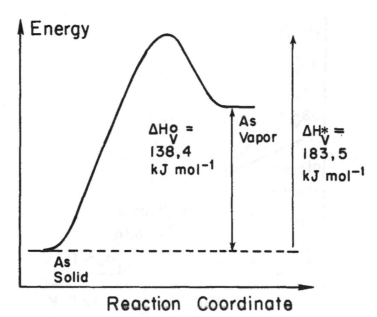

Fig. 2.12 Vaporization from the (111) face of an arsenic single crystal (Rosenblatt and Lee, 1968)

result was obtained by Davy and Somorjai (1971) in a study of the vaporization of single crystals of ice. According to their findings, at $T < 188$ K, $\Delta H^*_{\text{vap}} = \Delta H^o_{\text{vap}}$ with $\alpha_{\text{vap}} \simeq 1$, but at higher temperatures between 213 and 233 K, α_{vap} decreases steadily and ΔH^*_{vap} tends to $1/2 \Delta H^o_{\text{vap}}$. These data are illustrated on an Arrhenius type plot (Fig. 2.13). The value of p is a measure of the experimental rate of sublimation. Thus, a change in the slope of Fig. 2.13 indicates a change of mechanism of the phenomenon under study. Since at low temperatures $p = p_e$, it follows that $v_{\text{vap}} = v_{\text{vap,max}}$ and the vaporizing molecules are in equilibrium with the surface. The vaporization rate is then limited by the desorption of this molecule.

The desorption mechanism seems to involve a precursor state W_1:

$$W_2 \underset{k_{-1}}{\overset{k_1}{\rightleftharpoons}} W_1 \qquad \text{step 1}$$

$$W_1 \xrightarrow{k_2} W \qquad \text{step 2}$$

where W_1 and W_2 denote water molecules bound to the surface by one or two hydrogen bonds respectively, and W is a water molecule in the

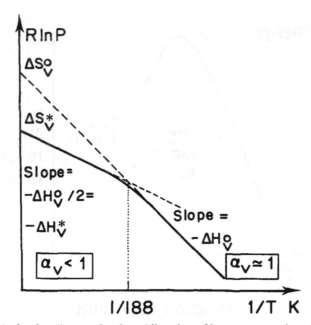

Fig. 2.13 Arrhenius diagram for the sublimation of ice: compensation effect (Davy and Somorjai, 1971)

vapor. Each step then is accompanied by the breaking (or making) of a hydrogen bond.

With a model of a stepped crystal, it can be supposed (Somorjai and Lester, 1967) that W_2 is an admolecule on the ledge whereas W_1 is a "walking" molecule on the terrace.

At low temperatures (LT), below 188 K, step (1) is equilibrated, and the rate determining step is the desorption:

$$v_{LT} = k_2[W_1] = k_2 \frac{k_1[W_2]}{k_{-1}} \qquad (2.4.11)$$

In the absence of any kinetic activation barrier, the expected change in enthalpy is that corresponding to the breaking of two hydrogen bonds ($W_2 \rightarrow W_1 \rightarrow W$). Hence $\Delta H^o_{vap} = 2\Delta H_B = \Delta H^*_{vap}$ where ΔH_B is the enthalpy change corresponding to the rupture of a single hydrogen bond.

At high temperatures (HT), above 200 K, step (1) becomes irreversible, and W is formed at the same rate as W_1. The observed rate is:

$$v_{HT} = k_1[W_2] \qquad (2.4.12)$$

and is not influenced by the second step. Hence the activation energy corresponds to the rupture of only one hydrogen bond. Hence:

$$\Delta H^*_{vap} = \Delta H_B = \tfrac{1}{2}\Delta H^o_{vap}$$

The compensation effect is due here to the existence of two successive steps. This kind of situation is typified by the "broken" Arrhenius diagram of Fig. 2.13 with a lower slope at the higher temperature.

2.5 BIMOLECULAR SURFACE REACTIONS

2.51 Collision Theory

In three dimensions (3D), (Moore, 1962, Boudart, 1968, Barrow, 1973), the collision theory of reaction rates is based on the rate of collisions between molecules A and B:

$$v_{3D} = \pi\sigma^2 \bar{v}_{3D,\mu}\{\exp(-E/RT)\}[A][B] = k[A][B] \qquad (2.5.1)$$

where E is the energy required for a collision to lead to reaction.

A probability factor P accounts for the fact that not all collisions lead to reaction (Boudart, 1968).

Consider the pre-exponential factor:

$$_AA_B = \pi\sigma^2\bar{v}_{3D,\mu} \quad (cm^3\ s^{-1}) \tag{2.5.2}$$

where $\sigma = 1/2\ (\sigma_A + \sigma_B)$, the mean molecular diameter, σ_A and σ_B being the diameters of molecules A and B, and $\bar{v}_{3D,\mu}$ being the mean molecular speed

$$\bar{v}_{3D,\mu} = (8kT/\pi\mu)^{\ddagger} \tag{2.5.3}$$

where $\mu = m_Am_B/(m_A + m_B)$ is the reduced mass based on the molecular masses of A and B.

In the case of associative desorption, collision theory deals with a two-dimensional space where the reaction takes place, with $[A]$ and $[B]$ being surface concentrations. By analogy with the formula in three dimensions, it appears intuitively that we could write:

$$_AA_A = \sigma\bar{v}_{2D} \quad (cm^2\ s^{-1}) \tag{2.5.4}$$

for the collision between two identical particles A and A. It says that the factor $_AA_A$ is the product of the mean molecular diameter and of the relative velocity. This value will be compared to that obtained from transition state theory. As an example, consider hydrogen at 550 K: $\sigma = 3 \times 10^{-8}$ cm, $\bar{v}_{2D} = 1.2 \times 10^4$ cm s^{-1}, and $_AA_A = 3.6 \times 10^{-4}$ cm^2 s^{-1}.

For desorption of hydrogen from various single crystals, Christmann collected values of the pre-exponential factor collected in Table 2.2. All

TABLE 2.2 Experimental values for the pre-exponential factor of the rate constant for associative desorption of H_2 (Christmann, 1979)

Metal	$_AA_A\ (cm^2\ s^{-1})$
Ni (100)	8.0×10^{-2}
	2.5×10^{-1}
	3.0
Ni (111)	2.0×10^{-1}
	2.3×10^{-2}
Pd (100)	1.0×10^{-2}
Pd (111)	1.3×10^{-1}
Ru (0001)	4.0×10^{-3}
Pt (111)	3.0×10^{-9}
Pt (997)	6.0×10^{-8}

results are larger than the above value except for two very anomalous low values on Pt.

To explain the higher experimental values, an explanation based on the work of Schmidt (1974) and Krylov et al. (1972) considers $\bar{v}_{2D}$ as a velocity of surface diffusion:

$$_A A'_A = \sigma \bar{v}' \qquad (2.5.5)$$

Following the ideas of Volmer, surface diffusion takes place by hopping between potential energy wells over potential barriers E_D. The distance between wells is λ (Fig. 2.14). If τ is the time between two successive jumps:

$$\bar{v}' = \lambda/\tau \quad (\text{cm s}^{-1}) \qquad (2.5.6)$$

and τ depends exponentially on temperature:

$$\tau = \tau_o \exp(E_D/RT) \qquad (2.5.7)$$

where τ_o is of the order of 10^{-13} s, the inverse of a normal pre-exponential factor of a unimolecular reaction. As to the value of λ, it should be close to that of σ. Hence:

$$_A A'_A = \sigma \bar{v}'$$
$$= \sigma^2 \tau_o^{-1} \exp(-E_D/RT) \qquad (2.5.8)$$

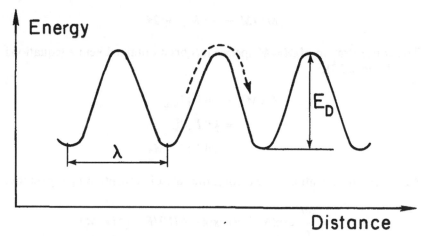

Fig. 2.14 Hopping surface diffusion (Volmer)

What is measured experimentally is the rate constant for desorption:

$$k_d = A_d \exp(-E_d/RT)$$

where:

$$A_d \equiv {}_A A'_A$$

Hence:

$$k_d = \sigma^2 \tau_o^{-1} \exp[-(E_D + E_d)/RT] \qquad (2.5.9)$$

This is yet another illustration of the compensation effect, related this time to surface diffusion. Indeed, equation (2.5.9) has higher values of both the pre-exponential factor ($\sigma^2 \tau_o^{-1} \simeq 10^{-2}$ cm^2 s^{-1}) and activation energy ($E_D + E_d$) than those based on collision theory ($\simeq 10^{-3}$ cm^2 s^{-1} and E_D respectively).

Another explanation of high values of the pre-exponential factors for bimolecular desorption is provided by transition state theory.

2.52 Transition State Theory

By surface diffusion, adsorbed species migrate:

$$\underset{****}{M \quad M} \longrightarrow \underset{****}{MM}$$

and associative desorption takes place from adjacent atoms:

$$M**M \longrightarrow M_2 + 2*$$

The concentration of $M**M$ species has been obtained before (equations 2.2.27 and 2.2.31):

$$[M**M] = \tfrac{1}{2}z[L]\theta_{MM}$$
$$= \tfrac{1}{2}z[L]\theta_M^2$$
$$= \tfrac{1}{2}z[L]^{-1}[M*]^2$$

According to transition state theory, the rate of desorption v_d is given by:

$$v_d = \frac{kT}{h} \exp(\Delta S^{o\dagger}/R) \exp(-\Delta H^{o\dagger}/RT)[M**M]$$

or:

$$v_d = \frac{kT}{h} \frac{1}{2}z[L]^{-1} \exp(\Delta S^{o\dagger}/R) \exp(-\Delta H^{o\dagger}/RT)[M*]^2 \quad (2.5.10)$$

Suppose first that $\Delta S^{o\dagger} = 0$. As an example take the (100) face of a bcc crystal for which $z = 4$ and $[L] = 5 \times 10^{14}$ cm^{-2}. At 550 K, the pre-exponential factor of (2.5.10) is:

$$A_d = 4.6 \times 10^{-2} \text{ cm}^2 \text{ s}^{-1}$$

which is close to several values of Table 2.2. This value is in perfect agreement with the experimental result of Tamm and Schmidt (1969) for a second-order desorption of hydrogen from the (100) face of bcc tungsten (Fig. 2.11):

$$2H* \longrightarrow H_2 + 2*$$

A peak temperature of 550 K was observed with a corresponding peak area of 2.5×10^{14} molecules of H_2 per cm^2. The reported experimental value of A_d was 4.2×10^{-2} cm^2 s^{-1}.

Values of A_d in Table 2.2 that are superior to about 5×10^{-2} cm^2 s^{-1} can be accounted for by assuming a positive value of $\Delta S^{o\dagger}$. In particular, if the transition state already possesses one rotational degree of freedom, $\exp(\Delta S^{o\dagger}/R)$ is of the order of 10. On the other hand, the anomalous low values of A_d in Table 2.2 are so much out of line with expected value that the experiments from which they were obtained must be considered as suspect. This position appears at least as reasonable as that of casting suspicion on transition state theory. Why would the latter explain desorption of hydrogen from Pd (111) but fail by eight orders of magnitude for Pt (111)?

If now there exists an attractive $(+)$ or repulsive $(-)$ interaction between adsorbed species:

$$E = E^o \pm 2z[\omega]\theta \quad (2.5.11)$$

as assumed by Goymour and King (1973), the rate of desorption becomes:

$$v_d = A_d \exp[-(E^o \pm 2z|\omega|\theta)/RT][A]^2$$

or more simply:

$$v_d = (v_d)_{\theta=0} \exp(\text{const} \times \theta) \quad (2.5.12)$$

as proposed by Langmuir as early as 1932. Thus for repulsive dipolar interactions, e.g., CO on W, the first case treated by Goymour and King, v_d decreases as θ decreases. By contrast, in the case of large adsorbed molecules, attractive dipolar interactions may lead to an increase in v_d as θ decreases (see §2.6).

All results obtained in this section are also applicable to bimolecular elementary steps between adsorbed species, the steps said in catalysis to be of the Langmuir-Hinshelwood type (see §1.34).

In temperature programmed desorption, three cases can occur for the desorption of adsorbed molecules. First, the molecule desorbs as such. Second, it desorbs after rupture or reaction in the adsorbed phase. Third, desorption takes place in the presence of a gas capable of reacting with the adsorbed molecule. The last two variations of the method can be called temperature programmed reaction.

2.6 TEMPERATURE PROGRAMMED REACTION

A first example deals with the destructive desorption of a pre-adsorbed reactant, formic acid, from the (110) plane of a nickel single crystal (McCarty et al., 1973). The corresponding thermal desorption spectrum is shown in Fig. 2.15.

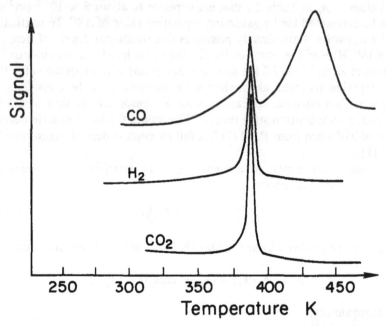

Fig. 2.15 Temperature programmed desorption of formic acid preadsorbed at 325 K on Ni (110). Heating rate 10 K s^{-1} (McCarty et al., 1973)

Besides desorption of water at low temperature (not shown on Fig. 2.15), three desorption peaks can be seen for CO, CO_2, and H_2, with the same asymmetric shape and the same peak temperature T_M (388 K). A further desorption peak for CO occurs at 438 K. When H_2 and CO are adsorbed separately and then desorbed, their desorption peaks appear at 353 and 438 K respectively.

The interpretation of these results involves a condensation of two formic acid molecules with loss of water to form the surface anhydride:

It is this species which is believed to undergo an auto-catalytic decomposition-desorption with a peak at 388 K and the observed products H_2, CO, and CO_2. It was verified that decreasing the initial surface coverage leads to an increase in the rate of destructive desorption as a result of attractive dipolar interactions between adsorbed anhydride molecules. This leads to an auto-catalytic phenomenon characterized by anomalously narrow desorption peaks at 388 K.

A second example deals with associative desorption and surface reaction (Madix, 1979b). The case is that of temperature programmed reaction of HCHO adsorbed on Ni (110). Changing surface coverage θ changes the nature of the desorbed products. With low θ, the only observed desorption peaks are those of H_2 and CO. At higher θ, CH_3OH appears. Two possibilities are considered. The first one involves a dissociative adsorption of formaldehyde at 200 K

$$3* + H_2CO \longrightarrow 2H* + CO*$$

and methanol formation according to:

$$H_2CO* + 2H* \longrightarrow CH_3OH + 3*$$

This possibility is rejected, as pre-adsorption of a mixture of D_2 and H_2CO should yield CH_2DOD which is not observed.

A second possibility involves intermolecular transfer of hydrogen.

$$2H_2CO* \longrightarrow CH_3OH* + CO*$$

TABLE 2.3   Temperature programmed reaction of H_2CO adsorbed at 200 K on Ni (110) (Madix, 1979b)

Product	E (kcal mol^{-1})	Pre-exponential factor s^{-1}	Relative yield
H_2	20	1.1×10^{14}	0.50
CO	33	8.5×10^{15}	1.00
CH_3OH	18	8.4×10^{14}	0.40
H_2O	18	7.3×10^{12}	0.08
CO_2	26	1.5×10^{15}	0.04
H_2CO	14	1.0×10^{13}	0.30

which agrees with experimental findings. Table 2.3 shows the rate constants determined in this study of temperature programmed reaction. Note in particular the compensation effect exhibited by the pre-exponential factors and activation energies. In particular, the values for CO and H_2CO are the highest and lowest respectively. Note also that several of the pre-exponential factors are greater than the "normal value," 10^{13} s^{-1}, indicating again a positive $\Delta S^{o\dagger}$.

This last remark must be kept in mind as the main results of this chapter are collected in Table 2.4. Whereas these order-of-magnitude estimates must be considered critically in the light of the discussions of this chapter, they are exceedingly useful in recognizing very abnormal behavior which may well be attributed to an experimental artifact. At any rate, the results of this chapter form the basis for the study of overall catalytic reactions which will be tackled in the subsequent chapter.

TABLE 2.4   Order of magnitude estimates of pre-exponential factors A for three surface elementary processes

Equation	Step	Value of A
$v_{a,o} = s_0 \dfrac{\bar{v}}{4} [M]$	adsorption	10^4 cm s^{-1}
$v_d = A \exp(-E/RT)[M*]$	unimolecular desorption or reaction	10^{13} s^{-1}
$v_{L.H.} = A \exp(-E/RT)[M*]^2$	bimolecular desorption or Langmuir-Hinshelwood elementary step	10^{-2} cm^2 s^{-1}

REFERENCES

Anderson, J. R. and Kemball, C. 1954. *Proc. Roy. Soc.* A223:361.

Balooch, M., Cardillo, M. J., Miller, D. R., and Stickney, R. E. 1974. *Surf. Sci.* 46:358.

Barrow, G. M. 1973. *Physical Chemistry.* New York: McGraw-Hill. 3rd ed. (1976), pp. 436 and 465.

Bell, A. T. and Hecker, W. C. 1981. *Anal. Chem.* 53:817.

Boudart, M. 1961. *Chem. Eng. Progress* 57:33.

Boudart, M. 1968. *Kinetics of Chemical Processes.* Englewood Cliffs, N.J.: Prentice-Hall, p. 47.

Boudart, M. 1975. In *Physical Chemistry*, ed. H. Eyring et al., 7:362. New York: Academic Press.

Bozso, F., Ertl, G., Grunze, M., and Weiss, M. 1977. *J. Catal.* 49:18.

Bozso, F., Ertl, G., and Weiss, M. 1977. *J. Catal.* 50:519.

Christmann, K. 1979. *Bull. Soc. Chim. Belge* 88:519.

Cremer, E. 1955. *Advan. Catal. Relat. Subj.* 7:75.

Cvetanovic, R. J. and Amenomiya, Y. 1967. *Advan. Catal. Relat. Subj.* 17:103.

Cvetanovic, R. J. and Amenomiya, Y. 1972. *Catal. Rev.* 6:21.

Davy, J. G. and Somorjai, G. A. 1971. *J. Chem. Phys.* 55:3624.

Ehrlich, G. 1961. *J. Appl. Phys.* 32:4.

Ehrlich, G. 1963. *Advan. Catal. Relat. Subj.* 14:271.

Ehrlich, G. and Stolt, K. 1980. *Ann. Rev. Phys. Chem.* 31:603.

Goymour, C. G. and King, D. A. 1973. *J. Chem. Soc. Faraday I* 69:739 and 749.

Grabke, H. J. 1967. *Ber. Bunsenges* 71:1067.

Hill, T. L. 1960. *Introduction to Statistical Thermodynamics.* Reading, Mass.: Addison-Wesley, p. 238.

Hirth, J. P. and Pound, G. M. 1963. *Condensation and Evaporation, Nucleation and Growth Kinetics.* Oxford: Pergamon.

Kemball, C. 1953. *Proc. Roy. Soc.* A217:376.

King, D. A. 1979. In *Chemistry and Physics of Solid Surfaces*, ed. R. Vanselow, 2:87. Boca Raton, Fla.: CRC Press.

Kossel, W. 1957. *Nach. Ges. Wiss.* (Göttingen), p. 135.

Krylov, O. V., Kislyuk, M. U., Shub, B. R., Gezalov, A. A., Maksimova, N. D., and Rufov, Yu N. 1972. *Kinet. i Katal.* 13:598.

Langmuir, I. 1932. *J. Am. Chem. Soc.* 54:2798.

Lennard-Jones, J. E. 1932. *Trans. Faraday Soc.* 28:333.

McCarty, J., Falconer, J., and Madix, R. J. 1973. *J. Catal.* 30:235.

Madix, R. J. 1979a. In *Chemistry and Physics of Solid Surfaces*, ed. R. Vanselov, 2:63. Boca Raton, Fla.: CRC Press.

Madix, R. J. 1979b. "The Physics of Surfaces Aspects of the Kinetics and Dynamics of Surface Reactions." Proceedings A.I.P. Conference, La Jolla Institute.

Moore, W. J. 1962. *Physical Chemistry.* 3rd ed. Englewood Cliffs, N.J.: Prentice-Hall, p. 237.

Mortensen, E. A. and Eyring, H. 1960. *J. Phys. Chem.* 64:846.

Petermann, L. A. 1972. *Progr. Surf. Sci.* 3:1.

Redhead, P. A. 1962. *Vacuum* 12:203.

Rosenblatt, G. M. and Lee, P. K. 1968. *J. Chem. Phys.* 49:2995.

Schmidt, L. D. 1974. *Catal. Rev.* 9:115.

Somorjai, G. A. 1981. *Chemistry in Two Dimensions, Surfaces.* Ithaca, N.Y.: Cornell University Press.

Somorjai, G. A. and Lester, J. E. 1967. *Progr. Solid State Chem.* 4:1.

Stranski, I. N. 1928. *Z. Phys. Chem.* 136:259.

Stranski, I. N. 1931. *Z. Phys. Chem.* 11 Abstract D:342.

Tamm, P. W. and Schmidt, L. D. 1969. *J. Chem. Phys.* 51:5352.

Taylor, H. S. 1931. *J. Am. Chem. Soc.* 53:518.

Tompkins, F. C. 1979. In *Chemistry and Physics of Solid Surfaces,* ed. R. Vanselow, 2:5. Boca Raton, Fla.: CRC Press.

Van Hardeveld, R. and Hartog, F. 1969. *Surface Sci.* 15:189.

Vol'kenshtein, F. F. 1949. *Zhur. Fiz. Khim.* 23:917.

Volmer, M. and Estermann, I. 1921. *Z. Phys.* 7:1.

Weinberg, W. H. and Merrill, R. P. 1971. *J. Vac. Sci. Technol.* 8:718.

Yates, J. T. and Madey, T. E. 1971. *Surface Sci.* 28:437.

Chapter 3

KINETICS OF
OVERALL REACTIONS

3.1 INTRODUCTION

The value of pre-exponential factors of rate constants can be used to assess the authenticity or validity of presumed elementary steps. Nothing much can be said a priori about activation energies. For adsorption equilibrium constants $K = \exp(\Delta S_a^o/R) \exp(-\Delta H_a^o/RT)$, the rule is that ΔH_a^o should be normally negative, as adsorption is in general exothermic, while entropy normally decreases upon adsorption, although the loss cannot exceed what was available to the fluid phase molecule, namely S_{fluid}^o (Vannice et al., 1979):

$$0 < -\Delta S_a^o < S_{fluid}^o$$

Yet endothermic adsorption and a gain of entropy upon adsorption are conceivable. In the latter case, imagine dissociative chemisorption of H_2 to freely mobile adsorbed atoms with two translation degrees of freedom each: the net gain of a translational degree of freedom could offset the loss of the rotational degrees of freedom of the free molecule. Nevertheless, the simple rules cited above are very useful.

Consider now the sequence of elementary steps for an overall reaction. The rate expression for an overall reaction can be obtained readily with two simplifications. First, it is assumed that all surface sites are identical both thermodynamically and kinetically, as befits a uniform surface. Second, it is assumed that there are no interactions between adsorbed species so that concentrations can be used instead of thermodynamic activities. Then classical rate equations for the overall reactions can be obtained in terms of the rate constants or equilibrium constants of the component elementary steps. Later on, the above simplifications will be relaxed.

3.2 KINETICS OF SINGLE REACTIONS: UNIFORM SURFACES

3.21 Single Path Reactions: General Equations (Horiuti, 1957; Temkin, 1973; Wagner, 1970; Happel, 1972)

a) Quasi-stationary state approximation. Consider a sequence of steps in which the active intermediate that is produced in a given step is a reactant in the next step. The catalytic sequence is closed in the sense that the active intermediate consumed in the first step is regenerated in the last. The rate of the overall reaction proceeding along the single path of the sequence may then be obtained. It is then assumed, first, that the active intermediates are conserved, i.e., the catalyst does not lose activity; second, that the stationary state is reached; and third, that active intermediates do not go directly from and to the catalyst and the fluid phase, as is the case of a single path reaction.

1. General relation between overall rate and rates of elementary steps. Suppose a sequence with n elementary steps, each one with a stoichiometric number σ_i. Following Temkin (1971), we can always write an identity which is readily verified and is valid whatever the number and the ordering of the steps:

$$(v_1 - v_{-1})v_2 v_3 \cdots v_n + v_{-1}(v_2 - v_{-2})v_3 v_4 \cdots v_n$$
$$+ v_{-1}v_{-2}(v_3 - v_{-3}) \cdots v_n$$
$$+ \cdots + v_{-1}v_{-2}v_{-3} \cdots (v_n - v_{-n})$$
$$= v_1 v_2 v_3 \cdots v_n - v_{-1}v_{-2}v_{-3} \cdots v_{-n} \quad (3.2.1)$$

With the quasi-stationary approximation, the net rate of the overall reaction v is equal to the difference between forward and reverse rates of each step:

$$\sigma_i v = v_i - v_{-i} \quad (3.2.2)$$

with the suitable weighing of the stoichiometric coefficient of that step (see §1.64).

Substitution of (3.2.2) into (3.2.1) yields a general relation of Temkin:

$$v = \frac{\prod_{i=1}^{n} v_i - \prod_{i=1}^{n} v_{-i}}{\sigma_1 v_2 \cdots v_n + v_{-1}\sigma_2 \cdots v_n + \cdots + v_{-1}v_{-2} \cdots v_{-(n-1)}\sigma_n} \quad (3.2.3)$$

But the net rate of the overall reaction can be expressed as the difference between a forward and a reverse rate:

$$v = \vec{v} - \overleftarrow{v} \quad (3.2.4)$$

Hence, by identification with equation (3.2.3):

$$\vec{v} = \frac{\prod\limits_{i=1}^{n} v_i}{D} ; \qquad \overleftarrow{v} = \frac{\prod\limits_{i=1}^{n} v_{-i}}{D} \tag{3.2.5}$$

when D is the denominator in equation (3.2.3). The result is:

$$\boxed{\frac{\vec{v}}{\overleftarrow{v}} = \frac{\prod\limits_{i=1}^{n} v_i}{\prod\limits_{i=1}^{n} v_{-i}}} \tag{3.2.6}$$

2. General relations for the elementary steps and for the overall reaction:

i) Relations with the affinity. Consider any elementary step:

$$A + B \underset{\overleftarrow{k}}{\overset{\overrightarrow{k}}{\rightleftharpoons}} C + D$$

The net rate v_i of a step i is:

$$v_i = v_i - v_{-i} \tag{3.2.7}$$

$$= \overrightarrow{k}[A][B] - \overleftarrow{k}[C][D]$$

$$= v_i \left(1 - \frac{[C][D]}{K[A][B]} \right) \tag{3.2.8}$$

with

$$K = \overrightarrow{k}/\overleftarrow{k} \tag{3.2.9}$$

which is always true for an elementary step that can be treated by transition state theory.

The affinity of the elementary step, i.e., its Gibbs free energy with the minus sign, is written as:

$$A_i = A_i^o - RT \ln \frac{[C][D]}{[A][B]} \tag{3.2.10}$$

when

$$A_i^o = RT \ln K \tag{3.2.11}$$

is the standard affinity. It follows that:

$$A_i = RT \ln K \frac{[A][B]}{[C][D]} \qquad (3.2.12)$$

or

$$\frac{K[A][B]}{[C][D]} = \exp(A_i/RT) \qquad (3.2.13)$$

Equation (3.2.8) then becomes:

$$v_i = v_i - v_{-i} = v_i[1 - \exp(-A_i/RT)] \qquad (3.2.14)$$

Replacing K (eq. 3.2.9) into (3.2.12) then yields:

$$A_i = RT \ln \frac{\vec{k}[A][B]}{\overleftarrow{k}[C][D]} = RT \ln \frac{v_i}{v_{-i}} \qquad (3.2.15)$$

or finally

$$\boxed{\frac{v_i}{v_{-i}} = \exp \frac{A_i}{RT}} \qquad (3.2.16)$$

The analogous relation for the overall reaction leads to the concept of rate determining step. The affinity of the overall reaction is:

$$A = \sum_i \sigma_i A_i$$

Equation (3.2.15) then gives:

$$A = RT \sum_i \ln \left(\frac{v_i}{v_{-i}}\right)^{\sigma_i}$$

or

$$A = RT \ln \prod_i \left(\frac{v_i}{v_{-i}}\right)^{\sigma_i} \qquad (3.2.17)$$

As shown schematically on Fig. 3.1, if there exists a rate determining step (rds), all other steps being equilibrated or quasi-equilibrated ($A_i = 0$,

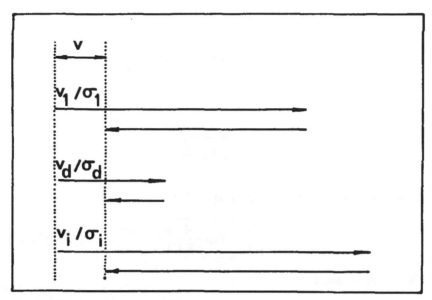

Fig. 3.1 Net rates (Tamaru, 1978)

$v_i = -v_i$ except for $i = d$), the above equation becomes:

$$A = \sigma_d A_d = RT \ln\left(\frac{v_d}{v_{-d}}\right)^{\sigma_d}$$

which becomes finally, since $\vec{v}/\overleftarrow{v} = v_d/v_{-d}$(eq. 3.2.6):

$$\sigma_d = \frac{A}{RT \ln\dfrac{\vec{v}}{\overleftarrow{v}}} \qquad (3.2.18)$$

Since A can be obtained from thermodynamic data, σ_d can be determined if only $\vec{v}$ and $\overleftarrow{v}$ are known.

In general, an average stoichiometric number $\bar{\sigma}$ can be defined, as any other average quantity, by means of:

$$\bar{\sigma} = \frac{\sum\limits_i \sigma_i A_i}{\sum\limits_i A_i} \qquad (3.2.19)$$

Since

$$\sum_i A_i = RT \ln \prod_i \frac{v_i}{v_{-i}} = RT \ln \frac{\vec{v}}{\overleftarrow{v}}$$

it follows that

$$\bar{\sigma} = \frac{A}{RT \ln \dfrac{\vec{v}}{\overleftarrow{v}}} \tag{3.2.20}$$

or in a form analogous to that of eq. (3.2.16):

$$\boxed{\frac{\vec{v}}{\overleftarrow{v}} = \exp \frac{A}{\bar{\sigma} RT}} \tag{3.2.21}$$

On the other hand:

$$v = \vec{v} - \overleftarrow{v} = \vec{v}\left(1 - \frac{\overleftarrow{v}}{\vec{v}}\right)$$

and

$$\boxed{v = \vec{v}\left[1 - \exp\left(-\frac{A}{\bar{\sigma} RT}\right)\right]} \tag{3.2.22}$$

Note that the definition of $\bar{\sigma}$ (eqs. 3.2.19 and 3.2.20) leads to indeterminate values when the system reaches equilibrium since all affinities become equal to zero and $\vec{v}$ is equal to $\overleftarrow{v}$. It will be shown on pp. 85–86 that $\bar{\sigma}$ preserves its physical meaning at equilibrium, as does the concept of a rate determining step.

   ii) Equilibrium constants and rate constants of an overall reaction. For an elementary step:

$$K = \vec{k}/\overleftarrow{k}$$

For an overall reaction, the forward and reverse rates can generally be expressed by the relations (Boudart, 1976):

$$\vec{v} = \vec{k} \prod_i [C_i]^{\alpha_i}/\phi \tag{3.2.23}$$

$$\overleftarrow{v} = \overleftarrow{k} \prod_i [C_i]^{\alpha_i}/\phi \tag{3.2.24}$$

C_i being the concentration of component i, and $\vec{\alpha}_i$, $\overleftarrow{\alpha}_i$ the orders of reaction with respect to that component. Orders may be positive, negative, or zero, with integral or fractional values. The function ϕ, such as the polynomial of eq. (3.2.5), depends on temperature and concentrations. Dividing side by side we get:

$$\frac{\vec{v}}{\overleftarrow{v}} = \frac{\vec{k}}{\overleftarrow{k}} \prod_i [C_i]^{\vec{\alpha}_i - \overleftarrow{\alpha}_i} = \exp \frac{A}{\bar{\sigma} RT} \qquad (3.2.25)$$

The affinity A is given by:

$$A = A^o - RT \ln \prod_i [C_i]^{v_i} = RT \ln K \prod_i [C_i]^{-v_i} \qquad (3.2.26)$$

where A^o is the standard value of the affinity and v_i is the stoichiometric coefficient of component i, positive for a product, negative for a reactant. Replacing A by its value (3.2.26) into (3.2.25) we get:

$$\frac{\vec{k}}{\overleftarrow{k}} \prod_i [C_i]^{\vec{\alpha}_i - \overleftarrow{\alpha}_i} = \left[k \prod_i [C_i]^{-v_i} \right]^{1/\bar{\sigma}} \qquad (3.2.27)$$

which gives after rearrangement:

$$\frac{\vec{k}}{\overleftarrow{k}} (K)^{-1/\bar{\sigma}} = \prod_i [C_i]^{\vec{\alpha}_i - \overleftarrow{\alpha}_i - (v_i/\bar{\sigma})} = \text{Constant} \qquad (3.2.28)$$

For this relation to be verified for all values of the concentrations, the exponents of all concentrations must be equal to zero, so that the constant must be equal to unity. Hence:

$$\boxed{\frac{\vec{k}}{\overleftarrow{k}} = K^{1/\bar{\sigma}}} \qquad (3.2.29)$$

This is a general relation which has been written under different forms. But here the exponent of K receives a physical meaning. The limitations are that $\bar{\sigma}$ and the orders α_i are assumed to be temperature independent and that the system is thermodynamically ideal.

The determination of $\bar{\sigma}$, or of σ_d if there exists a rate determining step, is possible by means of tracers, at equilibrium or away from equilibrium. The underlying principles have been formulated by Horiuti (1957), Temkin (1973), and Happel (1972).

b) Exchange rates or rates at equilibrium. The difficulty concerning the meaning of $\bar{\sigma}$ at equilibrium vanishes if one introduces exchange rates or rates at equilibrium as defined by Wagner (1970).

1. The exchange rate v_i^o of a step i is defined as the absolute value of the rate of the process, forward or reverse at equilibrium.

For any elementary step:

$$v_i = v_i - v_{-i} = v_i[1 - \exp(-A_i/RT)]$$

Near equilibrium, A_i tends to zero, v_i tends to v_{-i} and v_i^o. Since $|A_i/RT| \ll 1$, the exponential can be expanded with retention of the linear term only:

$$\exp(-A_i/RT)_{A_i \to 0} = 1 - A_i/RT \qquad (3.2.30)$$

Hence:

$$\boxed{v_i = v_i^o \frac{A_i}{RT}}_{|A_i/RT| \ll 1} \qquad (3.2.31)$$

Clearly, at equilibrium, the net rate v_i is zero but v_i^o is not zero. The definition of v_i^o is:

$$\boxed{v_i^o = \left[\frac{\partial(v_i - v_{-i})}{\partial(A_i/RT)}\right]_{A_i \to 0}} \qquad (3.2.32)$$

An analogous definition is written for the exchange rate v^o of the overall reaction:

$$\boxed{v^o = \left[\frac{\partial v}{\partial(A/\bar{\sigma}RT)}\right]_{A \to 0}} \qquad (3.2.33)$$

Thus, near equilibrium:

$$v_i = \sigma_i v = v_i - v_{-i} = v_i^o \frac{A_i}{RT} \qquad (3.2.34)$$

Summing up and taking into account the definition of $\bar{\sigma}$ (eq. 3.2.19), we get:

$$\sum_i \frac{\sigma_i}{v_i^o} v = \frac{\sum_i A_i}{RT}$$

or

$$v = \frac{1}{\sum_i \dfrac{\sigma_i}{v_i^o}} \frac{A}{\bar{\sigma} RT} \qquad (3.2.35)$$

But from the definition of v^o (eq. 3.2.33):

$$v^o = \frac{1}{\sum_i \dfrac{\sigma_i}{v_i^o}} \qquad (3.2.36)$$

which leads to the relation of Horiuti-Temkin:

$$\boxed{v = v^o \frac{A}{\bar{\sigma} RT}}\Bigg|_{|A/RT| \ll 1} \qquad (3.2.37)$$

which gives the overall rate v near equilibrium in terms of the exchange rate and the affinity. If v and v^o are determined, $\bar{\sigma}$ can be obtained.

2. Definitions of the rate determining step and of $\bar{\sigma}$ at equilibrium. If all the exchange rates v_i^o ($i \neq d$) are larger than v_d^o:

$$\sigma_d / v_d^o \gg \sigma_i / v_i^o \qquad (3.2.38)$$

equation (3.2.36) gives:

$$v_d^o \doteq v_d v^n \qquad (3.2.39)$$

which defines the rate determining step at equilibrium. On the other hand, from the definition of $\bar{\sigma}$ (eq. 3.2.19) and the relation:

$$\sigma_i v = v_i - v_{-i} = v_i^o \frac{A_i}{RT} \quad \text{when} \quad A_i \to 0$$

it follows that:

$$A_i = \sigma_i \frac{v}{v_i^o} RT \qquad (3.2.40)$$

and we get:

$$\boxed{\bar{\sigma} = \left(\sum_i \sigma_i^2/v_i^o\right)\Big/\left(\sum_i \sigma_i/v_i^o\right)}\Bigg|_{A_i \to 0} \qquad (3.2.41)$$

This shows that $\bar{\sigma}$ preserves its meaning at equilibrium (Boudart, 1975). In particular, if the inequalities (3.2.38) are satisfied, there exists a rate determining step with stoichiometric number σ_d, such that $\bar{\sigma} = \sigma_d$.

Thus if there exists a rate determining step, the value of $\bar{\sigma}$ is that of σ_d. Since the identification of the rate determining step is a fundamental

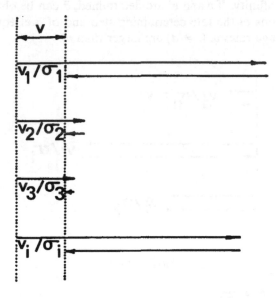

problem in catalysis, let us see how $\bar{\sigma}$ can be determined. It must be stressed, of course, that there are many cases, e.g., if all steps are irreversible, when a rate determining step does not exist.

c) Identification of the rate determining step.

1. Rate measurements near equilibrium. A first way to determine $\bar{\sigma}$ consists in measuring the rate near and at equilibrium (Horiuti and Nakamura, 1967):

$$v = \frac{v^o}{\bar{\sigma}} \frac{A}{RT} \quad \text{(eq. 3.2.37)}$$

Measuring v near equilibrium gives $v^o/\bar{\sigma}$ while the exchange rate v_o determined by isotopic tracing then yields $\bar{\sigma}$ or σ_d if there is a rate determining step. The measurement of v near equilibrium is particularly simple since, as indicated by eq. 3.2.37, the rate near equilibrium is proportional to any quantity that measures the distance away from equilibrium (see Boudart, 1968).

2. General method (Temkin, 1971, and Happel, 1972). As an example, consider ammonia synthesis on an iron catalyst. The tracer is ^{15}N. The postulated elementary steps of the sequence are as follows:

step					σ_i
1	N_2	$+\ 2*$	$\rightleftharpoons$	$2N*$	1
2	$N*$	$+\ H*$	$\rightleftharpoons$	$NH* + *$	2
3	$NH*$	$+\ H*$	$\rightleftharpoons$	$NH_2* + *$	2
4	NH_2*	$+\ H*$	$\rightleftharpoons$	$NH_3 + 2*$	2
5	H_2	$+\ 2*$	$\rightleftharpoons$	$2H*$	3
	N_2	$+\ 3H_2$	$=$	$2NH_3$	

The tracer goes through all steps except the last one, which is considered to be in quasi-equilibrium as witnessed by the very fast H_2-D_2 equilibration taking place under conditions of synthesis. For steps 1 to 4, we assume that equilibrium is not reached ($v_i \neq v_{-i}$). The overall reaction is:

$$^{15}NN + 3H_2 = {}^{15}NH_3 + NH_3$$

where the tracer ^{15}N is transferred from the reactant N_2 to the product NH_3 at a net rate:

$$\rho = \vec{\rho} - \overleftarrow{\rho} \qquad (3.2.42)$$

Furthermore:

$$v = \vec{v} - \overleftarrow{v} \qquad (3.2.43)$$

but ^{15}N does not go through all steps. Two sets of steps must be defined: those involving ^{15}N (set j), the other step not involving ^{15}N.

For $j = 1, 2, 3, 4$ the general Temkin relation is valid:

$$\frac{\vec{v}^{(j)}}{\overleftarrow{v}^{(j)}} = \frac{\prod_i^j v_i}{\prod_i^j v_{-i}} \tag{3.2.44}$$

The average stoichiometric number is:

$$\bar{\sigma}^{(j)} = \frac{\sum_{i=1}^{j} \sigma_i A_i}{\sum_{i=1}^{j} A_i} = \frac{A}{\sum_{i=1}^{j} A_i} \tag{3.2.45}$$

and according to equation (3.2.21):

$$\frac{\vec{v}^{(j)}}{\overleftarrow{v}^{(j)}} = \exp \frac{A}{\bar{\sigma}^{(j)} RT} \tag{3.2.46}$$

If x_A and x_B are the molar fractions of the tracer atom in the reactant N_2 and the product NH_3, the rates of exchange of the tracer are:

$$\vec{\rho} = \mu x_A \vec{v}^{(j)}$$
$$\overleftarrow{\rho} = \mu x_B \overleftarrow{v}^{(j)} \tag{3.2.47}$$

where μ is the number of equivalent atoms that can be labeled.

From equations (3.2.42), (3.2.43), and (3.2.47), we get:

$$\vec{v}^{(j)} = \frac{\rho - \mu x_B v}{\mu(x_A - x_B)}$$

$$\overleftarrow{v}^{(j)} = \frac{\rho - \mu x_A v}{\mu(x_A - x_B)} \tag{3.2.48}$$

Dividing the two previous equations side by side, we get:

$$\frac{\vec{v}^{(j)}}{\overleftarrow{v}^{(j)}} = \frac{\rho - \mu x_B v}{\rho - \mu x_A v} = \exp \frac{A}{\bar{\sigma}^{(j)} RT} \tag{3.2.49}$$

Hence:

$$\bar{\sigma}^{(j)} = \frac{A}{RT} \ln \frac{\rho - \mu x_A v}{\rho - \mu x_B v} \qquad (3.2.50)$$

Thus $\bar{\sigma}^{(j)}$ can be determined from the net overall rate v and from the net rate of tracer transfer ρ. If the steps through which the tracer does not pass are equilibrated, as in the example considered:

$$\bar{\sigma}^{(j)} = \bar{\sigma}$$

and if moreover there exists a rate-determining step, then:

$$\bar{\sigma} = \sigma_d = \bar{\sigma}^{(j)} \qquad (3.2.51)$$

In other words, the stoichiometric number determined from the transfer of a tracer is also that of the overall reaction and of the rate determining step. In the chosen example, σ_d can be equal to 1 or 2. Experimental studies have reported both values.

Horiuti et al. (1953, 1954) found $\sigma_d = 2$ for an iron catalyst near equilibrium. Tanaka (1965) found $\sigma_d = 2$ for synthesis and $\sigma_d = 1$ for decomposition, far from equilibrium. The value $\sigma_d = 2$ is compatible with step 2 being rate-determining. But most of the evidence indicates that adsorption of dinitrogen is the rate determining step, which means $\sigma_d = 1$ (de Boer, ed., 1960). Still, it cannot be said whether the adsorption is dissociative or not. Indeed, the following sequence can be envisaged:

step					σ_i
1	N_2	$+$ *	$\rightleftharpoons$	N_2*	1
2	N_2*	$+ H_2$	$\rightleftharpoons$	N_2H_2*	1
3	N_2H_2*	$+ H_2$	$\rightleftharpoons$	N_2H_4*	1
4	N_2H_4*	$+ H_2$	$\rightleftharpoons$	$2NH_3 +$ *	1
	N_2	$+ 3H_2$	$=$	$2NH_3$	

The first two steps are those that are considered, those by which the enzyme nitrogenase fixes dinitrogen. If so, σ_d would still be equal to unity. Nevertheless, the evidence for dissociative chemisorption of dinitrogen on iron is now overwhelming. The two sequences, dissociative on iron and associative on nitrogenase, might well be respective characteristics of industrial catalysis at high temperature and enzymatic catalysis at low temperature. To improve substantially the activity of present-day iron catalysts for ammonia synthesis, it might be necessary to modify radically the nature of the steps in the sequence on iron.

3.22 Rate Expression for the Overall Reaction: Two-Step Sequences

We have discussed the relation between the overall rate v and the average stoichiometric number which can lead to the identification of the rate determining step if there is one:

$$v = \vec{v}[1 - \exp(-A/\bar{\sigma}RT)]$$

Thus only the forward rate $\vec{v}$ needs to be known besides $\bar{\sigma}$ and the affinity A.

The knowledge of v and its explicit dependence on pressure, temperature, and composition serves two major purposes: the elucidation of reaction mechanism and the design of chemical reactors. To be sure, proof of a reaction mechanism requires other evidence besides kinetics. But even if imperfectly known, a mechanism is likely to yield a better rate expression which, in time, can be improved for optimization of reactor performance.

First let us obtain the rate equation for a two-step reaction. While these are rare, many multistep reactions can be treated as if they were sequences of two steps.

a) Sequence of two reversible elementary steps:

$$S_1 + A_1 \underset{k_{-1}}{\overset{k_1}{\rightleftarrows}} B_1 + S_2 \qquad (1)$$

$$\frac{S_2 + A_2 \underset{k_{-2}}{\overset{k_2}{\rightleftarrows}} B_2 + S_1}{A_1 + A_2 \quad = \quad B_1 + B_2} \qquad \begin{array}{l}(2)\\ \text{: overall reaction}\end{array}$$

S_1 and S_2 denote empty and occupied sites respectively, with:

$$[L] = [S_1] + [S_2] \qquad (3.2.52)$$

A classic example of such a sequence is the water gas shift reaction:

$$\frac{\begin{array}{l} * \quad + H_2O \rightleftharpoons H_2 \; + O* \\ O* \; + CO \; \rightleftharpoons CO_2 + * \end{array}}{H_2O + CO \quad = \quad CO_2 + H_2} \qquad \text{: overall reaction}$$

To simplify the writing, the following notations will be adopted:

$$a_1 = k_1[A_1] \qquad a_2 = k_2[A_2]$$
$$a_{-1} = k_{-1}[B_1] \qquad a_{-2} = k_{-2}[B_2]$$

(3.2.53)

where the a_i's may be considered as pseudo-first-order rate constants and the inverse of the mean lifetime of the Bodenstein active intermediates, in this case S_1 and S_2.

The relation of Temkin (eq. 3.2.3) leads to the overall rate:

$$v = \frac{v_1 v_2 - v_{-1} v_{-2}}{v_{-1} + v_2}$$
$$= \frac{[S_1](a_1 a_2 - a_{-1} a_{-2})}{a_{-1} + a_2}$$

(3.2.54)

But since the order of the steps in Temkin's relations does not matter, we also have:

$$v = \frac{[S_2](a_1 a_2 - a_{-1} a_{-2})}{a_1 + a_{-2}}$$

(3.2.55)

Dividing the two preceding equations side by side leads to the ratio u of the concentrations of empty and occupied sites:

$$u = \frac{[S_1]}{[S_2]} = \frac{a_{-1} + a_2}{a_1 + a_{-2}}$$

(3.2.56)

This ratio is readily obtained from the quasi-stationary state approximation, noting that $v = v_1 = v_2$. The use of the Temkin relations illustrate how they can lead to the elimination of the unknowns by rotation of the indices. Thus the rate can always be obtained, at least in principle, although it can be difficult when the rates of the elementary steps are not linear in the concentrations of the active intermediates, e.g., for dissociative adsorption (see Boudart, 1968). The turnover rate is then:

$$v_t = \frac{v}{[L]} = \frac{[S_1](a_1 a_2 - a_{-1} a_{-2})}{[S_1] + [S_2](a_2 + a_{-1})}$$

(3.2.57)

or finally:

$$v_t = \frac{v}{[L]} = \frac{a_1 a_2 - a_{-1} a_{-2}}{a_1 + a_{-1} + a_2 + a_{-2}} \qquad (3.2.58)$$

This can be generalized to a sequence of i steps: the numerator will still be of the form of the law of mass action while the denominator is a polynomial of i^2 terms. In practice, the least number of parameters is best and simplifications that correspond to various experimental situations are in order.

1. Case of a saturated surface. This condition means that $[S_2] \gg [S_1]$, which simplifies eq. (3.2.57) to:

$$v_t = \frac{k_1 k_2 [A_1][A_2] - k_{-1} k_{-2} [B_1][B_2]}{k_1[A_1] + k_{-2}[B_2]} \qquad (3.2.59)$$

where the a_i's have been replaced by their values.

If, besides, step (1) is irreversible, $k_{-1} = 0$, we get an equation which starts to resemble those of Langmuir, popularized by Hinshelwood:

$$v_t = \frac{k_2[A_2]}{1 + \dfrac{k_{-2}[B_2]}{k_1[A_1]}} \qquad (3.2.60)$$

2. Case of a rate determining step. If step (2) is rate-determining and irreversible, we have $a_2, a_{-2} \ll a_1, a_{-1}$ and v_t becomes in the absence of $B_1 ([B_1] = 1)$:

$$v_t = \frac{k_2 K_1 [A_2]}{1 + K_1[A_1]} \qquad (3.2.61)$$

with $K_1 = k_1/k_{-1}$. This is the classic equation of Michaelis-Menten in enzyme catalysis, Langmuir-Hinshelwood in physical chemistry, and Hougen-Watson in chemical engineering.

So far, it has been supposed, with Langmuir, that all catalytic sites are uniform and do not interact. This formalism has its limitations. But it frequently works, and the equations obtained can be compared to empirical power rate laws:

$$v = k[A_1]^{\alpha_1}[A_2]^{\alpha_2} \cdots \qquad (3.2.62)$$

where k is an apparent rate constant and α_1, α_2, ... are apparent orders with respect to reactants and products. Let us now examine cases where these orders are fractional.

b) Fractional orders. These are frequently reported. What do they mean?

Example 1. Catalytic decomposition of germane.

$$GeH_4(g) = Ge(s) + 2H_2(g)$$

This corresponds to the simplest case where a single reactant enters in the overall rate equation:

$$v = k[GeH_4]^{\alpha_1}$$

with $0 < \alpha_1 < 1$. Germane decomposes to yield H_2 and a mirror of germanium that keeps growing. At high temperature, Hogness and Johnson (1932) have reported $\alpha_1 = 1/3$.

But, as shown by Tamaru et al. (1955), this fractional order has no kinetic meaning. Indeed, the reaction consists in the superposition of two parallel ones: one taking place at the surface (zero order) and the second occurring homogeneously (first order):

$$v = k_1[GeH_4]^0 + k_2[GeH_4]^1 \qquad (3.2.63)$$

Example 2. Decomposition of stibine between 273 and 348 K. This led to the first case of heterogeneous kinetics reported in the literature. It was studied by Stock and Bodenstein (1907). The overall reaction:

$$SbH_3 = Sb + 3/2 H_2$$

obeys the rate expression

$$v = k[SbH_3]^{0.6} \qquad (3.2.64)$$

where the exponent depends neither on temperature nor on pressure, a result that remained unexplained by the original authors.

Later on, it was proposed (Laidler, 1965) that there are two elementary steps, the first one an equilibrated adsorption (with constant K_1) followed by a decomposition in the adsorbed phase (with constant k_2):

$$SbH_3 + * \underset{}{\overset{K_1}{\rightleftharpoons}} SbH_3* \qquad (1)$$

$$SbH_3* \quad \overset{k_2}{\longrightarrow} \cdots \qquad (2)$$

The symbol $\wedge$ denotes, following the Japanese school, a rate-determining step. The species SbH_3* constitutes what will be called the most abundant reaction intermediate (*mari*), all others being present at the surface at much inferior concentration levels. It must be noted that the *mari* is not necessarily the most abundant surface species, since the surface may be covered by species that do not take part in the reaction.

The rate is then that of the rate-determining step:

$$v = k_2[SbH_3*] \tag{3.2.65}$$

from which $[SbH_3*]$ can be eliminated by means of:

$$K_1 = [SbH_3*]/[SbH_3][*]$$

and

$$[L] = [*] + [SbH_3*]$$

the result is:

$$v = k_2[L] \frac{K_1[SbH_3]}{1 + K_1[SbH_3]} \tag{3.2.66}$$

The ratio $\theta = K_1[SbH_3]/(1 + K_1[SbH_3])$ is the fraction of the surface covered by the *mari*. This relation is the Langmuir adsorption isotherm, which can be written in an approximate way:

$$\theta = \text{Const} \times [SbH_3]^n$$

with $0 < n < 1$. Indeed, when $[SbH_3]$ is small enough (Fig. 3.2), $\theta = K[SbH_3]$ or $\propto [SbH_3]^1$, and when it is large enough, $\theta = 1$ or $\theta \propto [SbH_3]^0$. In between $\theta \propto [SbH_3]^n$ with $0 < n < 1$ and the equation (3.2.66) can be rewritten as:

$$v = k[L] \, \text{Const} \times [SbH_3]^n \tag{3.2.67}$$

with $0 < n < 1$ as found experimentally (eq. 3.2.64). But since the exponent is independent of temperature between 273 and 348 K, the heat of adsorption of SbH_3 should be equal to zero. If so, why would there be adsorption and reaction?

Let us consider an alternative sequence, consisting of an irreversible adsorption step:

$$SbH_3 + * \xrightarrow{k_1} \cdots \tag{1}$$

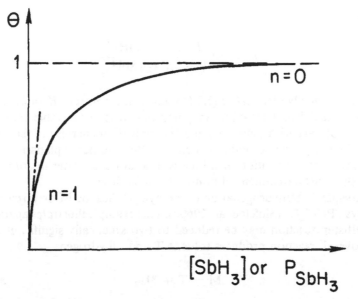

Fig. 3.2 Langmuir adsorption isotherm: fractional coverage θ versus pressure or concentration

on a surface with a *mari* consisting of SbH_2* which decomposes irreversibly:

$$SbH_2* \xrightarrow{k_2} \cdots \qquad (2)$$

At the quasi-stationary state:

$$v = v_1 = v_2$$

It follows that:

$$v = k_1[SbH_3][*] = k_2[SbH_2*] \qquad (3.2.68)$$

with:

$$[L] = [*] + [SbH_2*]$$

Thus:

$$[SbH_2*] = \frac{(k_1/k_2)[SbH_3][L]}{1 + (k_1/k_2)[SbH_3]} \qquad (3.2.69)$$

Hence:

$$v = [L] \frac{k_2(k_1/k_2)[SbH_3]}{1 + (k_1/k_2)[SbH_3]} \qquad (3.2.70)$$

which is formally identical to (3.2.66) but with a constant $K = k_1/k_2$ that has now a different meaning. In particular, it is conceivable that k_1/k_2 would not vary with temperature if the activation energies of the two irreversible steps were essentially identical. Yet another explanation of the kinetics of decomposition of stibine will be discussed later in connection with the kinetic treatment of non-uniform surfaces.

Example 3. Dehydrogenation of methylcyclohexane M on a reforming catalyst, Pt/Al_2O_3 (Sinfelt et al., 1960b). This example illustrates again how a multistep reaction may be reduced to two kinetically significant steps. The overall reaction produces toluene T and dihydrogen:

$$M = T + 3H_2 \qquad (3.2.71)$$

with $v = k[M]^n$ and $n = 0.5$.

Sufficiently far away from equilibrium, the rate equation may also be written in the form:

$$v = \frac{\alpha[M]}{1 + \beta[M]} \qquad (3.2.72)$$

where β changes with temperature. A discrimination between the two explanations invoked in the case of stibine necessitates other data. Let us present again these two explanations.

Case 1. Equilibrated adsorption of methylcyclohexane.

$$M + * \overset{K_1}{\rightleftharpoons} M* \qquad (1)$$

with M* as *mari*. The rds is then:

$$M* \overset{k_2}{\xrightarrow{\hspace{0.6cm}}} \cdots \qquad (2)$$

Hence:

$$v_2 = k_2[M*] = [L]k_2 \frac{K_1[M]}{1 + K_1[M]} \qquad (3.2.73)$$

Case 2. There are only two practically irreversible steps. The adsorption of M to an unspecified intermediate:

$$M + * \xrightarrow{\ k_1\ } \cdots \tag{1}$$

is followed by several unspecified steps leading to the formation of the *mari* T*:

The *mari* then desorbs:

$$T* \xrightarrow{\ k_2\ } * + T \tag{2}$$

Application of the quasi-steady state approximation leads again to:

$$v_2 = [L]k_2 \frac{(k_1/k_2)[\mathrm{M}]}{1 + (k_1/k_2)[\mathrm{M}]} \tag{3.2.74}$$

This explanation is preferred because separate experiments of Sinfelt et al. have shown that benzene added to the feed inhibits the rate only slightly. Yet it is expected from Case 1 that it would displace methylcyclohexane from the surface in a very competitive way. With Case 2, since step (2) is practically irreversible, the steady-state surface coverage by T* must be far above its equilibrium value. Thus benzene added to the feed will not compete favorably for sites with toluene since benzene will cover the surface only to the extent of its equilibrium amount.

Example 4. Decomposition of ammonia on a platinum filament. The overall reaction is:

$$2NH_3 = N_2 + 3H_2$$

The data are those of Löffler and Schmidt (1976). They cover an unusually large range of temperatures (573 to 1473 K) and ammonia pressures (10^{-3} to 10 torr). The overall rate exhibits an order with respect to ammonia going from zero at low temperature to unity at high temperature. The authors show that all the data can be represented by a classic rate expression (see note on p. 117):

$$v = [L]k_2 \frac{K_1[\mathrm{NH_3}]}{1 + K_1[\mathrm{NH_3}]} \tag{3.2.75}$$

which can be interpreted by the now familiar sequence:

$$NH_3 + * \underset{}{\overset{K_1}{\rightleftharpoons}} NH_3* \tag{1}$$

$$NH_3* \xrightarrow{k_2} \cdots \tag{2}$$

with NH_3* being the *mari*. The latter choice appears doubtful, as one would expect NH_3* to be weakly adsorbed as compared to other dissociated fragments such as $N*$, $NH*$, ... which would be more abundant than NH_3*, especially at high temperatures.

But the same form of rate equation can be obtained by retaining only two irreversible steps:

$$NH_3 + 2* \xrightarrow{k_1} NH_2* + H* \tag{1}$$

(other unspecified, kinetically
nonsignificant steps)

$$2N* \xrightarrow{k_2} N_2 + 2* \tag{2}$$

with $N*$ being the *mari*. The case is thus similar to those of decomposition of stibine and dehydrogenation of methylcyclohexane. In particular, application of the quasi-steady state approximation gives:

$$v = v_1 = v_2 = k_1[NH_3][*]^2[L]^{-1} = k_2[N*]^2[L]^{-1} \tag{3.2.76}$$

The justification of $[L]^{-1}$ is discussed in §2.23. Thus:

$$\frac{[N*]}{[*]} = \left(\frac{k_1}{k_2}[NH_3]\right)^{\frac{1}{2}} \tag{3.2.77}$$

and

$$[L] = [*] + [N*] = [N*]\left(1 + \frac{[*]}{[N*]}\right)$$

Hence:

$$[N*] = [L]\frac{(k_1/k_2[NH_3])^{\frac{1}{2}}}{1 + (k_1/k_2[NH_3])^{\frac{1}{2}}} \tag{3.2.78}$$

The overall rate is then:

$$v = k_2[L]\left[\frac{\{(k_1/k_2)[NH_3]\}^{\frac{1}{2}}}{1 + \{(k_1/k_2)[NH_3]\}^{\frac{1}{2}}}\right]^2 \tag{3.2.79}$$

Again, just from fitting the kinetic data, it is not possible to discriminate between equations (3.2.75) and (3.2.79), as they are so very similar. The mechanism described above, with irreversible adsorption of the reactant and irreversible desorption of the *mari*, explain very satisfactorily all the data of Tamaru et al. (1980) and of Boudart et al. (1982) for the high temperature and low pressure decomposition rate of ammonia on tungsten and molybdenum respectively, with simultaneous measurement of the surface concentration of N* by Auger electron spectroscopy. Disturbing the stationary state by flashing the metal catalyst, the rates of return to steady state of the two postulated irreversible steps of adsorption and desorption could be evaluated separately.

Consider another example of a sequence that can be brought down to two steps by postulating an rds and a *mari*.

Example 5. Synthesis of ammonia. The overall reaction:

$$N_2 + 3H_2 = 2NH_3$$

for which we have already considered a sequence of steps, can be treated as a two-step reaction. For simplicity's sake, the reaction will be assumed to proceed far from equilibrium so that the reverse rate can be neglected.

a) Dissociative adsorption of dinitrogen. The adsorption step is the rds that produces the *mari* N*:

$$N_2 + 2* \xrightarrow{k_1} 2N* \tag{1}$$

Because of this assumption, all the other equilibrated steps may be summed up into an overall equilibrated reaction:

$$N* + 3/2H_2 \underset{}{\overset{K_2}{\rightleftharpoons}} NH_3 + * \tag{2}$$

It must be noted that (2) is not an elementary step, as witnessed by the presence of a non-mechanistic stoichiometric coefficient of 3/2.

The overall rate is then:

$$v = k_1[N_2][*]^2[L]^{-1} \tag{3.2.80}$$

But as usual:

$$[L] = [*] + [N*]$$

and

$$K_2 = \frac{[N*][H_2]^{3/2}}{[NH_3][*]} \tag{3.2.81}$$

The equilibrium constant is written in such a way that it is large when the adsorption of N* is also large. It is easy to obtain:

$$[*] = \frac{[L]}{1 + \{K_2[NH_3]/[H_2]^{3/2}\}}$$ (3.2.82)

Hence:

$$v = [L]\frac{K_1[N_2]}{[1 + \{K_2[NH_3]/[H_2]^{3/2}\}]^2}$$ (3.2.83)

b) Associative desorption of dinitrogen. In this case, we write:

$$N_2 + * \xrightarrow{k_1} N_2*$$ (1)

$$N_2* + 3H_2 \overset{K_2}{\rightleftharpoons} 2NH_3 + *$$ (2)

The first step is still the rds creating the *mari* N$_2$*. Now we have:

$$v = k_1[N_2][*]$$ (3.2.84)

with the overall equilibrium constant:

$$K_2 = \frac{[N_2*][H_2]^3}{[*][NH_3]^2}$$ (3.2.85)

Again:

$$[L] = [N_2*] + [*]$$

Hence:

$$[*] = \frac{[L]}{1 + K_2\dfrac{[NH_3]^2}{[H_2]^3}}$$

Finally:

$$v = [L]\frac{k_1[N_2]}{1 + (K_2[NH_3]^2/[H_2]^3)}$$ (3.2.86)

It would be very hard to discriminate between equations (3.2.86) and (3.2.83) on the basis of goodness of fit of kinetic data. To decide whether the adsorption of dinitrogen is dissociative or not, other experiments are needed.

c) Dissociative adsorption of N_2 with NH* as most abundant reaction intermediate. The sequence considered is:

$$2* \quad + N_2 \xrightarrow{k_1} 2N* \tag{1}$$

$$2* \quad + H_2 \rightleftharpoons 2H* \tag{2}$$

$$N* \quad + H* \longrightarrow NH* + * \tag{3}$$

$$NH* + H_2 \overset{K_4}{\rightleftharpoons} NH_3 + * \tag{4}$$

Steps (1) and (3) are irreversible. Thus neither one can be considered as rds. The last reaction is not an elementary one but a sum of equilibrated steps. We write the three usual equations:

$$v = v_1 = k_1[*]^2[N_2][L]^{-1} \tag{3.2.87}$$

$$[L] = [*] + [NH*] \tag{3.2.88}$$

$$K_4 = \frac{[NH*][H_2]}{[NH_3][*]} \tag{3.2.89}$$

As before, K_4 is written in the direction of adsorption. We obtain:

$$[*] = \frac{[L]}{1 + (K_4[NH_3]/[H_2])} \tag{3.2.90}$$

and finally

$$v = [L]\frac{k_1[N_2]}{\{1 + (K_4[NH_3]/[H_2])\}^2} \tag{3.2.91}$$

The last expression is quite close to a preceding one (3.2.83). To decide whether NH* is the *mari*, one might use the isotope jump technique of Tamaru (1978). At the steady state, one would replace at time $t = 0$ the normal feed $N_2 + 3H_2$ by $N_2 + 3D_2$. Spectroscopic observation of the surface by infrared could detect the relaxation of NH* to ND*, as shown on Fig. 3.3.

Example 6. Hydrogenolysis of ethane. The overall reaction is:

$$C_2H_6 + H_2 = 2CH_4$$

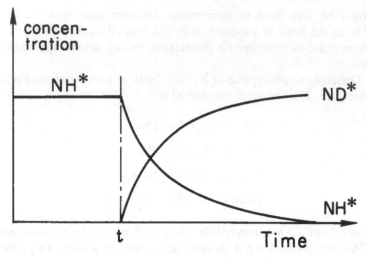

Fig. 3.3 Isotope jump technique following the method of Tamaru (1978)

On many metal catalysts, the reaction proceeds at a rate that is represented by a power rate law:

$$v = k[C_2H_6]^\alpha [H_2]^\beta \tag{3.2.92}$$

with $\alpha \simeq 1$ and $-2 < \beta < 0.5$. To explain these results, Cimino et al. (1954) postulated a two-step reaction sequence:

$$C_2H_6 \ + \ * \ \overset{K_1}{\rightleftharpoons} \ C_2H_x* \ + \ \frac{6-x}{2}\,H_2 \tag{1}$$

$$C_2H_x* \ + \ H_2 \ \overset{k_2}{\longrightarrow} \cdots \tag{2}$$

where the *mari* is the partially dehydrogenated species C_2H_x*. Since step (2) is the rds, what follows it, leading ultimately to methane, is not kinetically significant. The first reaction is not an elementary step but an overall equilibrium with an equlibrium constant:

$$K = \frac{[C_2H_x*][H_2]^{(6-x)/2}}{[C_2H_6][*]} \tag{3.2.93}$$

The usual balance of sites is:

$$[L] = [*] + [C_2H_x*] \tag{3.2.94}$$

and the rate is that of the rds:

$$v = k[C_2H_x*][H_2] \qquad (3.2.95)$$

With the usual substitutions, we get:

$$v = [L]k_2[H_2] \frac{K_1[C_2H_6][H_2]^{(x-6)/2}}{1 + K_1[C_2H_6][H_2]^{(x-6)/2}}$$

The ratio on the right-hand side is a generalized Langmuir adsorption isotherm that can be approximated by means of:

$$\text{const} \times ([C_2H_6]/[H_2]^{(6-x)/2})^n$$

with $0 < n < 1$. Hence:

$$v = k'[C_2H_6]^n[H_2]^{\left(1 - n\frac{6-x}{2}\right)} \qquad (3.2.96)$$

This is of the same form as the power rate law (3.2.92) found experimentally. This rate law was first verified in the case of a series of Fischer-Tropsch synthesis catalysts, which produce methane and higher hydrocarbons from CO and H_2.

Table 3.1 shows the experimental values of n, identical to the order α of eq. (3.2.92), as well as values of β and x. The calculated value of β, namely β_{calc}, is given by $1 - n(6 - x)/2$. Only integral values of x have been used.

Catalyst 1 is nickel, which does not produce higher hydrocarbons in Fischer-Tropsch synthesis but practically only CH_4. It is seen that the *mari* appears to be a completely dehydrogenated species ($x = 0$), a rather startling result. Catalysts 2, 3, and 4 are mediocre Fischer-Tropsch catalysts, but the last two, 5 and 6, are excellent ones. The latter correspond

TABLE 3.1 Hydrogenolysis of ethane $v = k[C_2H_6]^\alpha[H_2]^\beta$

Catalyst	$n_{exp} \equiv \alpha$	β_{exp}	x	β_{calc}
1 Ni	0.7	−1.2	0	−1.1
2 Fe	1.0	−0.7	2	−1.0
3 Fe + 0.05% K	0.9	−0.7	2	−0.8
4 Fe + 0.06% Li	0.8	−0.4	2	−0.6
5 Fe + 0.6% K	0.7	+0.3	4	+0.3
6 Fe + KOH	0.6	+0.1	4	+0.4

to the least dehydrogenated surface species. The former seem to involve an in-between degree of dehydrogenation at the surface. Note that the proposed rate equation accounts for negative and positive values of β.

In a previous study, Morikawa et al. (1937) had found a power rate law for the hydrogenolysis of propane C_3H_8 on nickel:

$$v = k[C_3H_8]^{0.9}[H_2]^{-2.6}$$

By a similar reasoning as for ethane hydrogenolysis, the very negative value of β can be calculated by means of $\beta_{calc} = 1 - n\dfrac{8-x}{2} = -2.6$, where $x = 0$ as before for ethane on nickel and $n = 0.9$, the observed order with respect to propane. In spite of this striking agreement, there are some objections to the simple mechanism proposed, and the matter will be discussed again later in connection with structure-sensitive reactions and kinetics on a non-uniform surface.

In conclusion, the above six examples permit us to generalize the treatment of single-path reactions as follows. Two major simplifications are introduced as a result of the assumption of a most abundant reaction intermediate, *mari*, and, if possible, of a rate determining step, rds. Three rules can then be formulated. First, if in a sequence the rds produces or destroys the *mari*, the sequence can be reduced to two steps, an adsorption equilibrium and the rds, all other steps having no kinetic significance (Examples 2 and 3, Cases 1 and 4).

Second, if all steps are practically irreversible and there exists a *mari*, only two steps need to be taken into account: the adsorption step and the reaction (or desorption) step of the *mari*. All other steps have no kinetic significance. In fact, they may be reversible, in part or in whole. Examples are 2, 3, and Cases 2 and 4.

Third, all equilibrated steps following an rds that produces the *mari* may be summed up in an overall equilibrium reaction. Examples are 5a and 5b. Similarly, all equilibrated steps that precede an rds that consumes the *mari* may be represented by a single overall equilibrium reaction (Example 6).

Two general procedures can be used. First, if the rds consumes the *mari*, the concentration of the latter is obtained from the equilibrium relation that is available. Second, if the steps of the two-step sequence are practically irreversible, the quasi-stationary state approximation leads to the solution.

The frequent occurrence of two-step reactions is helpful, in general and, as we shall see, in the application of the theory of non-uniform surface kinetics that is currently restricted to two-step sequences. But of course, simplifying the assumptions may break down as process variables so that a different rate equation must be adopted. An example is the oxidation of carbon monoxide on the (111) face of a single crystal of palladium, studied by Engel and Ertl (1978a):

$$CO + 1/2\,O_2 = CO_2$$

The authors have studied the reaction exhaustively with the help of low-energy electron diffraction, photoelectron spectroscopy, temperature programmed desorption and pulsed molecular beam relaxation spectroscopy. The nature, ordering, and binding energy of the adsorbed species were ascertained. It was checked that no oxygen dissolved in the metal and that the dissociation of O_2 took place but not that of CO (Conrad et al., 1977). The pressure range covered was not too wide ($10^{-8} < P_{CO,CO_2} < 10^{-6}$ torr), but the temperature interval was quite large: $400 < T < 700$ K The authors conclude on the basis of extensive data that under these conditions, the reaction takes place through an irreversible step of the Langmuir-Hinshelwood type, i.e., a reaction between adsorbed species, following a reversible adsorption step for CO and an irreversible adsorption step for O_2. It is further assumed that $[O*O] \ll [O*]$. The sequence is:

$$* \;\; + \;\text{CO} \underset{k_{-1}}{\overset{k_1}{\rightleftharpoons}} \text{CO}* \tag{1}$$

$$* \;\; + \;\; O_2 \xrightarrow{k_2} O*O \tag{2}$$

$$O*O + \;\; * \longrightarrow 2O* \tag{2'}$$

$$\text{CO}* + O* \xrightarrow{k_3} CO_2 + 2* \tag{3}$$

A mechanism of Eley-Rideal in which gaseous CO would react with adsorbed oxygen could be eliminated by the pulsed molecular beam data. Thus the overall rate should be given by:

$$v = k_3[O*][\text{CO}*]$$

But the data show that the values of fractional coverage, θ_O and θ_{CO} are not simple functions of P_{O_2} and P_{CO}. The rate law depends on temperature, partial pressures, co-adsorption of O_2 and CO, surface diffusion, and the structure of the adsorbed layer that depends on the surface coverage.

Fig. 3.4 shows how the overall rate v changes with temperature at the stationary rate. A maximum rate is reached at about 550 K for the partial

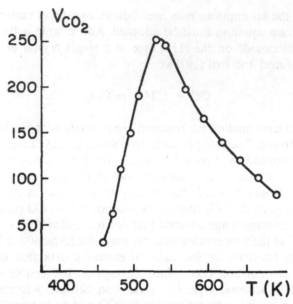

Fig. 3.4. Rate of formation of CO_2 versus temperature:
$P_{O_2} = 4 \times 10^{-7}$ torr; $P_{CO} = 1 \times 10^{-6}$ torr

pressures indicated. The interpretation of the kinetic results below and above 550 K goes as follows.

1. For $T < 550$ K and $P_{CO} \leq 10^{-6}$ torr. The rate goes up with temperature. The *mari* is CO∗ in this range, but its surface concentration decreases as T goes up and becomes quite small at 550 K (Ertl and Neumann, 1974). Thus the rate of adsorption of dioxygen is at first inhibited by CO but increases as temperature goes up. Thus the rate equation is given by:

$$v = k' \frac{[O_2]}{[CO]}$$

This can be seen as we write:

$$v = v_2 = k_2[O_2][*] \quad \text{(step 2)}$$

Furthermore, if we admit that the adsorption of CO is quasi-equilibrated:

$$\frac{k_1}{k_{-1}} = K_1 = \frac{[CO*]}{[CO][*]} = \text{Const} \times \exp(-\Delta H^o_{ads\,CO}/RT)$$

If we remember that $[O*O] \lll [O*] \ll [CO*]$:

$$[L] = [*] + [CO*]$$

The concentration of free sites is then:

$$[*] = \frac{[L]}{1 + K_1[CO]}$$

$$v = k_2[L] \frac{[O_2]}{(1 + K_1[CO])}$$

At high values of surface coverage, this becomes:

$$v = [L] \frac{k_2}{K_1} \frac{[O_2]}{[CO]} \qquad (3.2.97)$$

which is the observed rate law. Since $-\Delta H^o_{ads\,CO} = 32$ kcal mol^{-1} and the activation energy of step (2) is effectively zero, the rate increases with temperature with an apparent activation energy of 32 kcal mol^{-1}. It must be noted that this quantity is thermochemical in nature and not a kinetic quantity.

By separate relaxation measurements it was possible to measure the rate constant of the Langmuir-Hinshelwood step (3), to obtain its activation energy which was found to be equal to 25 kcal mol^{-1} and to determine its pre-exponential factor, which for a value of $[L] = 10^{15}$ cm^{-2} is equal to 1.1×10^{-3} cm^2 s^{-1} in agreement with the expected value (eq. 2.5.10). This result of Engel and Ertl (1978a, b) must be considered of particular importance, as it is the first time that the pre-exponential factor and the activation energy of an identified Langmuir-Hinshelwood surface step have been determined directly on a well-defined single crystal surface.

2. For $T > 550$ K, $P_{CO} \leq 10^{-6}$ torr. The surface coverage θ_{CO} is small, and nonstationary rate measurements show that for $\theta_O \geq 0.08$, v remains independent of θ_O (Engel and Ertl, 1978b). By low-energy electron diffraction, it was found by Conrad et al. (1978) that islands of oxygen are formed at the surface, with a *local* value of $\theta_O = 0.25$ inside the islands which consist of an ordered structure. The formation of CO_2 takes place at the periphery of these islands. Carbon monoxide molecules diffuse on the surface. If the diffusion path before CO desorbs is larger than the distance between islands, the rate is independent of θ_O. The

reaction still takes place in the adsorbed phase:

$$v = k_3[O*][CO*]$$

with $[O*] \simeq$ Const and $[L] \simeq [*]$. Hence:

$$[CO*] = [L]K_1[CO]$$

and

$$v = \text{Const} \times k_3[L]K_1[CO] \tag{3.2.99}$$

or simply

$$v = k'[CO]$$

which is the experimental result. The apparent activation energy is now equal to $\Delta H^o_{ads\,CO} + E_3$, a negative value $-32 + 25$ kcal mol^{-1}, so that the rate goes down as temperature goes up. One may also say that the apparent activation energy confirms independently the value $E_3 = 25$ kcal mol^{-1} determined by relaxation methods.

3.23 Reaction Networks

a) Parallel reactions. As an example, we shall consider the hydrogenation of aromatics on a nickel catalyst (Raney nickel), in the liquid phase. The reaction with a *single* aromatic compound A, is said to proceed in a two-step sequence:

$$A_1 + * \overset{K_1}{\underset{}{\rightleftharpoons}} A_1* \tag{1}$$
$$A_1* \overset{k_1}{\underset{}{\longrightarrow}} \cdots \tag{2}$$

Let us assume that A_1* is the *mari*. In the liquid phase, because of the high concentration of reactants, the surface will be practically saturated with the *mari*, so that:

$$[A_1*] = [L]$$

Then:

$$v_1 = k_1[L] \tag{3.2.101}$$

hence the zero order reaction rate as observed by Wauquier and Jungers (1957).

If now two aromatic molecules A_1 and A_2 are present in solution, the competition of these two for site occupancy leads to a very different result. For each molecule, the preceding sequence remains valid, but the competition for the same sites means that, in the case of saturation as before:

$$[L] = [*] + [A_1*] + [A_2*] = [A_1*] + [A_2*] \qquad (3.2.102)$$

For the first reactant:

$$v_1 = k_1[A_1*]$$

$$[A_1]K_1 = \frac{[A_1*]}{[*]}$$

while for A_2 the adsorption equilibrium constant is:

$$[A_2]K_2 = \frac{[A_2*]}{[*]}$$

From the last two equations, we get by summing them up side by side:

$$[*] = \frac{(L)}{K_1[A_1] + K_2[A_2]}$$

hence:

$$v_1 = k_1[L] \frac{K_1[A_1]}{K_1[A_1] + K_2[A_2]} \qquad (3.2.103)$$

and correspondingly:

$$v_2 = k_2[L] \frac{K_2[A_2]}{K_1[A_1] + K_2[A_2]} \qquad (3.2.104)$$

The last two relations give an interesting result. At constant volume, we can write:

$$v_1 = -\frac{d[A_1]}{dt} \quad \text{and} \quad v_2 = -\frac{d[A_2]}{dt}$$

Hence:

$$\frac{d[A_1]}{d[A_2]} = \frac{k_1 K_1 [A_1]}{k_2 K_2 [A_2]} \qquad (3.2.105)$$

Integrating between $[A_i]_o$ and $[A_i]_t$, concentration values at time zero and time t, we get:

$$\ln \frac{[A_1]_o}{[A_1]_t} = \frac{k_1 K_1}{k_2 K_2} \ln \frac{[A_2]_o}{[A_2]_t} \qquad (3.2.106)$$

The slope of the straight line corresponding to eq. (3.2.106) and shown in Fig. 3.5 is equal to the ratio $k_1 K_1/k_2 K_2$ which is a selectivity constant $S_{1,2}$ for the competing processes considered.

One can expect to find a simple result if one takes in succession couples of competing reactants and determines separately the selectivity constant for each couple. Starting and finishing with the same reactant, in the first

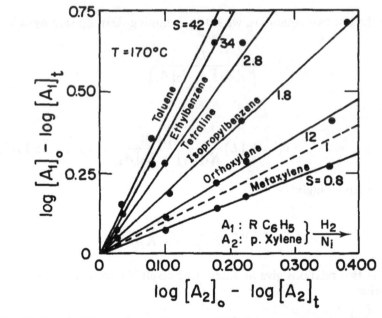

Fig. 3.5 Data of Wauquier and Jungers (1957) for the co-hydrogenation of A_1 (any aromatic) and A_2 (p-xylene) on Raney nickel at 170°C

and in the last pair, one expects:

$$S_{1,2} \times S_{2,3} \times \cdots \times S_{i,1} = 1 \qquad (3.2.107)$$

This result has been verified by Wauquier and Jungers (1957) for the co-hydrogenation of aromatics on nickel. It remains one of the most convincing arguments in favor of the self-consistency of kinetic parameters based on the theory of uniform surfaces.

Another interesting result of the study of these parallel reactions is due to the interplay between kinetic and thermodynamic factors entering in the selectivity constant. Thus, if for a pair of co-reactants, it happens that $k_1 < k_2$ with $K_1 > K_2$, the total rate of reaction is expected to increase with time as sketched on Fig. 3.6, as the least reactive ingredient grabs more of the surface sites at first, but relinquishes them to the more active reactant as its own concentration decreases.

A third result in competitive liquid phase hydrogenation is the effect of the solvent on selectivity. This effect was demonstrated by Wauquier and Jungers in the case of the co-hydrogenation of cyclohexene (subscript 1) and acetone (subscript 2) on nickel. It was first established that if these compounds are hydrogenated separately, the zero order rate constants k_1 and k_2 do not change from solvent by solvent. Since the selectivity constant $S_{1,2}$ was found to depend on the nature of the solvent, it is con-

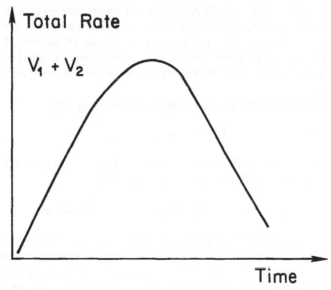

Fig. 3.6 Variation of the total rate with time when $k_1 < k_2$ and $K_1 > K_2$

cluded that the solvent affects the values of K_1 and K_2, as the solvent changes the partition coefficients of the co-reactants between the liquid phase and the surface.

The changes in the measured selectivity constant can be understood very simply. Thus isopropyl alcohol, a polar solvent, is hydrogen-bonded with acetone so that non-polar cyclohexene is rejected to the surface, and $S_{1,2}$ is higher than that found in a non-polar solvent cyclohexane. In the latter case, cyclohexene and cyclohexane tend to form an ideal solution while acetone is rejected to the surface with a lowering of $S_{1,2}$.

The next step in this work consisted in affecting the surface itself by adsorption of modifiers consisting of polar (e.g., pyridine) or non-polar molecules (e.g., stearic acid). In the first case, acetone will be drawn to the surface with a lowering of $S_{1,2}$. In the second case, the long hydrocarbon chain of the fatty acid provides a surface medium that favors the adsorption of cyclohexene with an increase in $S_{1,2}$.

By combining the effect of the solvent and the effect of the modifiers, Wauquier and Jungers succeeded in changing $S_{1,2}$ by a factor as large as 2,000 in the co-hydrogenation of cyclohexene and acetone. As happens so frequently in catalysis, a control of selectivity is achieved at the price of activity. Indeed, the modifiers occupy many surface sites that become unavailable to the reaction. A similar control of selectivity toward asymmetric hydrogenation has been achieved by adsorbing on Raney nickel optically active modifiers (Izumi and Tai, 1977).

Thus the very simple mechanisms proposed in this section not only account for the observed kinetic behavior, but also lead to the prediction and the explanation of new phenomena, including those related to the most subtle and challenging problem of catalysis, namely, selectivity. The double role of kinetic and thermodynamic factors in selectivity is accounted for.

b) Consecutive reactions. The simplest network is now:

$$A_1 \Longrightarrow A_2 \Longrightarrow A_3$$

A first example deals with the selective hydrogenation of acetylenic compounds on palladium (Meyer and Burwell, 1963):

$$\text{alkyne} + \text{H}_2 \xrightarrow{k_1, K_1} \text{alkene} + \text{H}_2 \xrightarrow{k_2, K_2} \text{alkane}$$

In this case, because K_1 is much larger than K_2, the reaction is very selective, and it is possible to stop the reaction after a practically quantitative transformation of the alkyne into the alkene, as the latter gets access to the surface only after the former has almost completely disappeared (see Table 3.2).

TABLE 3.2 Mole fraction of the products of reaction between
2-butyne and deuterium at 287 K on palladium (Meyer and Burwell, 1963)

2-butyne	$CH_3 - C \equiv C - CH_3$	0.220
cis-2-Butene-2,3-d_2	$\overset{\displaystyle CH_3}{\diagdown} C = C \overset{\displaystyle CH_3}{\diagup}$ $\quad$ D $\qquad$ D	0.772
trans-2-butene-2,3-d_2	D $\qquad$ CH_3 $\diagdown C = C \diagup$ $CH_3 \qquad$ D	0.007
1-Butene	$CH_2 {=} CH - CH_2 - CH_3$	0.000
Butane	$CH_3 - CH_2 - CH_2 - CH_3$	0.001

Other networks involve so-called *rake* mechanisms (Germain, 1969).

$$A_1 \qquad A_2 \qquad A_i$$
$$\updownarrow \qquad \updownarrow \qquad \updownarrow$$
$$A_1* \rightleftharpoons A_2* \cdots\cdots A_i*$$

A second example is the hydrogenolysis of cyclohexylamine on a platinum evaporated film (Kemball and Moss, 1960). The products of the reaction are ammonia, benzene, and cyclohexane. The time evolution of the composition of the reaction mixture is shown on Fig. 3.7. In an overall manner:

$$C_6H_{11}NH_2 \overset{H_2}{\Longrightarrow} C_6H_6; \; NH_3 \overset{H_2}{\Longrightarrow} C_6H_{12}$$

The surprising result is that benzene is not only formed at all, but that it is formed in substantial quantities with further hydrogenation to cyclohexane only after some cyclohexylamine has reacted. Yet under the conditions of the work, the equilibrium benzene $+ 3H_2 =$ cyclohexane is displaced completely to the right. It appears that after adsorption of cyclohexylamine, the C—N bond is hydrogenolyzed with formation of ammonia. But the hydrocarbon ring desorbs as benzene rather than as cyclohexane until later in the reaction. This kinetic behavior is of course not in contradiction with thermodynamic imperative, but illustrates the fact that *during the course of a reaction*, stable or reactive intermediates may be formed in quantities exceeding those predicted by equilibrium *at the end of the reaction.*

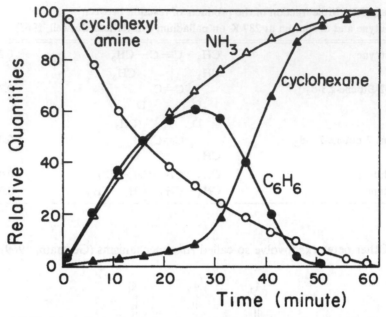

Fig. 3.7 Hydrogenolysis of cyclohexylamine in the presence of excess dihydrogen on a platinum film at 461 K (Kemball and Moss, 1960)

c) Bifunctional reactions. The next example is one of the reactions in catalytic reforming of hydrocarbons: the isomerization of n-pentane (n-C_5) to iso-pentane (i-C_5) on a Pt/Al_2O_3 catalyst (Sinfelt et al., 1960a).

$$n\text{-}C_5 = i\text{-}C_5$$

with an observed power rate law:

$$v = k\left\{\frac{[n\text{-}C_5]}{[H_2]}\right\}^\alpha$$

$\alpha = 0.5$. Although dihydrogen does not appear in the stoichiometric equation for reaction, it is present in large quantities in the reacting mixture. Its inhibiting effect on the rate is somewhat puzzling at first.

The explanation lies in the bifunctionality of Pt/Al_2O_3 with a (de)hydrogenating function of the metal and an acidic function of the alumina. The latter accounts for the isomerization of alkenes.

The mechanism then involves the formation on platinum of an intermediate, n-pentene $C_5^=$ which migrates to the acidic sites of the support where it undergoes isomerization, the rate-determining step:

$$\text{n-C}_5 \xrightleftharpoons{\text{Pt},K} \text{n-C}_5^= + \text{H}_2 \tag{1}$$

$$* + \text{n-C}_5^= \xrightleftharpoons{K_1} \text{n-C}_5^= * \tag{2}$$

$$\text{n-C}_5^= * \xrightarrow{k_1,\,A} \text{i-C}_5^= * \tag{3}$$

$$\text{i-C}_5^= * \xrightleftharpoons[\text{Pt}]{} \text{i-C}_5^= + * \tag{4}$$

$$\text{H}_2 + \text{i-C}_5^= \xrightleftharpoons{} \text{i-C}_5 \tag{5}$$

Acid sites on Al_2O_3 are denoted by $*$. The isopentene $\text{i-C}_5^=$ desorbs from the Al_2O_3, migrates back to Pt where it is hydrogenated to the final product.

The overall rate, far from equilibrium, is:

$$v = k_1 [\text{n-C}_5^= *]$$

Expressing $[\text{n-C}_5^= *]$ in terms of K_1, we get:

$$v = [L]k_1 \frac{K_1 [\text{n-C}_5^=]}{1 + K_1 [\text{n-C}_5^=]} \cong k' [\text{n-C}_5^=]^\alpha \tag{3.2.108}$$

where $0 < \alpha < 1$.

Since n-pentene is in equilibrium with n-pentane and hydrogen, thanks to the metallic function:

$$[\text{n-C}_5^=] = K[\text{n-C}_5]/[\text{H}_2]$$

and the rate becomes finally:

$$v \simeq k' \left[K \frac{[\text{n-C}_5]}{[\text{H}_2]} \right]^\alpha \equiv k'' \left[\frac{[\text{n-C}_5]}{[\text{H}_2]} \right]^\alpha \tag{3.2.109}$$

in agreement with the experimental result. The approximation introduced in eq. (3.2.108) is not essential, as will be seen later in connection with the treatment of non-uniform surface kinetics which gives to the exponent α an exact meaning.

Further support in favor of the proposed mechanism was provided by Sinfelt et al. (1960a). As shown on Fig. 3.8, the data represented by circles correspond to calculated partial pressures of pentene in equilibrium with pentane and dihydrogen. They are rate data obtained during the isomerization of n-pentane on $\text{Pt}/\text{Al}_2\text{O}_3$, and the slope of the line in Fig. 3.8 is

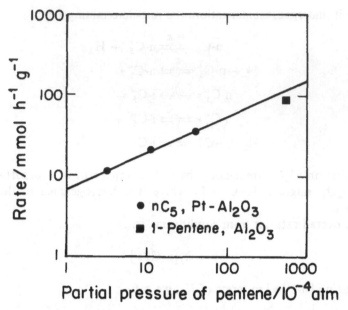

Fig. 3.8 Rate of isomerization versus partial pressure of pentene. The slope gives the observed reaction order $\alpha = 0.5$ for the isomerization of pentene (Sinfelt et al., 1960a).

one-half as expected from eq. (3.2.108). The experimental rate represented by a square is that of n-pentene isomerization on Al_2O_3 alone without platinum. In spite of the bold extrapolation, this rate on Al_2O_3 is very near the line corresponding to the data on Pt/Al_2O_3. This is an elegant demonstration of bifunctional behavior. It must be noted, however, that direct isomerization of alkanes on metals such as platinum can also take place at other rates under quite different conditions, as shown by Gault and his school (Garin and Gault, 1975). But these results do not affect the discussion of bifunctional isomerization given above.

REFERENCES

Boer, J. H. de, ed. 1960. *The Mechanism of Heterogeneous Catalysis.* Amsterdam: Elsevier.

Boudart, M. 1968. *Kinetics of Chemical Processes.* Englewood Cliffs, N.J.: Prentice-Hall.

Boudart, M. 1975. In *Physical Chemistry,* ed. H. Eyring, 7:350. New York: Academic Press.

Boudart, M. 1976. *J. Phys. Chem.* 26:2879.

Boudart, M., Egawa, S., Oyama, S. T., and Tamaru, K. 1982. *J. Chim. Phys.* 78:987.

Cimino, A., Boudart, M., and Taylor, H. 1954. *J. Phys. Chem.* 58:796.

Conrad, H., Ertl, G., Kuppers, J., and Latta, E. E. 1977. *Surf. Sci.* 65:245.

Conrad, H., Ertl, G., and Kuppers, J. 1978. *Surf. Sci.* 76:323.

Engel, T. and Ertl, G. 1978a, *Chem. Phys. Lett.* 54:95.

Engel, T. and Ertl, G. 1978b, *J. Chem. Phys.* 69:1267.

Ertl, G. and Neumann, M. 1974. *Z. Phys. Chem.* (N.F.) 90:127.

Garin, F. and Gault, F. G. 1975. *J. Am. Chem. Soc.* 97:4466.

Germain, J. E. 1969. *Catalytic Conversion of Hydrocarbons.* New York: Academic Press, p. 259.

Happel, J. 1972. *Catal. Rev.* 6:221.

Hogness, T. R. and Johnson, W. C. 1932. *J. Am. Chem. Soc.* 54:3583.

Horiuti, J. 1957. *J. Res. Inst. Catal.* (Hokkaido Univ.) 5:1.

Horiuti, J. et al. 1953. *J. Res. Inst. Catal.* (Hokkaido Univ.) 2:87.

Horiuti, J. et al. 1954. *J. Res. Inst. Catal.* (Hokkaido Univ.) 3:185.

Horiuti, J. and Nakamura, T. 1967. *Advan. Catal. Relat. Subj.* 17:1.

Izumi, Y. and Tai, A. 1977. *Stereo-differentiating Reactions.* Tokyo: Kodansha, and New York: Academic Press.

Kemball, C. and Moss, R. L. 1960. *Trans. Faraday Soc.* 56:154.

Laidler, K. J. 1965. *Chemical Kinetics.* New York: McGraw-Hill, p. 273.

Löffler, D. G. and Schmidt, L. D. 1976. *J. Catal.* 41:440.

Meyer, E. F. and Burwell, R. L., Jr. 1963. *J. Am. Chem. Soc.* 85:2877.

Morikawa, K., Trenner, N., and Taylor, H. S. 1937. *J. Am. Chem. Soc.* 59:1103.

Sinfelt, J. H., Hurwitz, H., and Rohrer, J. C. 1960a. *J. Phys. Chem.* 64:892.

Sinfelt, J. H., Hurwitz, H., and Shulman, R. A. 1960b. *J. Phys. Chem.* 64:1559.

Stock, A. and Bodenstein, M. 1907. *Ber.* 40:570.

Tamaru, K. 1978. *Dynamic Heterogeneous Catalysis.* London: Academic Press.

Tamaru, K., Boudart, M., and Taylor, H. S. 1955. *J. Phys. Chem.* 59:801.

Tamaru, K., Shindo, H., Egawa, C., and Onishi, T. 1980. *J. C. S. Faraday I* 2:280.

Tanaka, K. 1965. *J. Res. Inst. Catal.* (Hokkaido Univ.) 13:119.

Temkin, M. I. 1971. *Int. Chem. Eng.* 11:709.

Temkin, M. I. 1973. *Ann. N.Y. Acad. Sci.* 213:79.

Vannice, M. A., Hyun, S. H., Kalpacki, B., and Liauh, W. C. 1979. *J. Catal.* 56:358.

Wagner, C. 1970. *Advan. Catal. Relat. Subj.* 21:323.

Wauquier, J. P. and Jungers, J. C. 1957, *Bull. Soc. Chim.*, p. 1280.

NOTE: According to a private communication of Professor Löffler, his extensive data on the decomposition of ammonia on platinum are fitted better by his equation (3.2.75) than by our equation (3.2.79). However, as shown by R. W. McCabe (*J. Catal.* 79 [1983]:445) in the case of ammonia decomposition on nickel, an equation formally identical to the equation of Löffler and Schmidt can be obtained by the mechanism suggested on p. 98, consisting of two irreversible steps, the second one being first order in the concentration of adsorbed nitrogen instead of second order as assumed on p. 98. Thus, as we conclude at the top of p. 99, just from fitting the kinetic data for ammonia decomposition on platinum, it is not possible to discriminate between the two mechanisms presented on p. 98.

Chapter 4

KINETICS OF TWO-STEP REACTIONS ON NON-UNIFORM SURFACES

4.1 INTRODUCTION

In the preceding chapter, general results on the kinetics of surface-catalyzed reactions have been obtained with the usual simplification of a Langmuir surface with catalytic sites of uniform thermodynamic and kinetic properties. This restriction will now be relaxed for the case of two-step reactions, which were shown to be much more widespread than might be anticipated, thanks to the frequent existence of a rate-determining step and of a most-abundant-reaction intermediate.

Beyond the theory of Temkin (1957, 1965), the foundation of which was discussed by Khammouma (1972), we will consider the problem of the optimization of catalysts by simply replacing a non-uniform surface by a collection of catalysts with different properties.

4.2 STUDY OF TWO-STEP SINGLE-PATH REACTIONS

4.21 Fundamental Definitions

The model of the non-uniform surface is that of a collection of ensembles E_j (Fig. 4.1) of catalytic sites with identical thermodynamic and kinetic properties in each ensemble. Each ensemble contains dS'_j sites per cm^2. The total site density $[L]$ is obtained by integration over all ensembles:

$$\int dS'_j = [L] \tag{4.2.1}$$

That an integration is carried out rather than a summation means a continuous site distribution. The justification is in the trend of binding energies with coverage for adsorption of a given molecule on various crystallographic planes of a catalytic solid. An example for the adsorption of H_2 on W (Fig. 4.2; Schmidt, 1974) shows that it is easy to smooth out

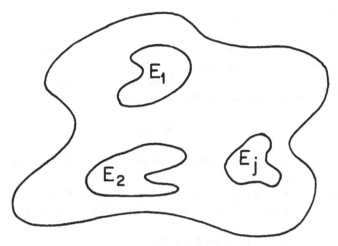

Fig. 4.1 Non-uniform surface

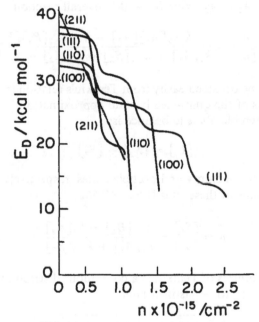

Fig. 4.2 Adsorption of hydrogen on tungsten. Values of the activation energy of desorption from various planes (Schmidt, 1974) compared to the smoothed-out measured heats of adsorption on a polycrystalline wire (data of Roberts, 1935, see Miller, 1949).

the curves for individual planes into a single continuous one, as observed in fact for a polycrystalline tungsten surface.

The total rate is, then, the sum of the partial rates on each ensemble. With the assumption of a continuous distribution, we write:

$$v = \int v_{t,j} \, dS'_j \tag{4.2.2}$$

where $v_{t,j}$ is the turnover rate on sites of the ensemble E_j, calculated for a uniform surface.

The problem is to carry out the integration above. For each ensemble, we can write the value of $v_{t,j}$ for a two-step catalytic sequence:

$$S_1 + A_1 \underset{k_{-1}}{\overset{k_1}{\rightleftharpoons}} B_1 + S_2 \tag{1}$$

$$S_2 + A_2 \underset{k_{-2}}{\overset{k_2}{\rightleftharpoons}} B_2 + S_1 \tag{2}$$

$$\overline{A_1 + A_2 \overset{K}{=\!=\!=} B_1 + B_2} \quad \text{overall reaction}$$

$$v_{t,j} = \frac{v_j}{[L_j]} = \frac{k_1 k_2 [A_1][A_2] - k_{-1} k_{-2} [B_1][B_2]}{k_1[A_1] + k_{-2}[B_2] + k_2[A_2] + k_{-1}[B_1]} \tag{4.2.3}$$

This result can be obtained easily from Temkin's relation (eq. 3.2.3) or directly by means of the quasi-steady state approximation.

For each ensemble, the site balance is:

$$[L_j] = [S_1] + [S_2] \tag{4.2.4}$$

where S_1 and S_2 are empty and occupied sites, respectively. The ratio of the concentrations of these sites is (eq. 3.2.56):

$$u = \frac{[S_1]}{[S_2]} = \frac{k_{-1}[B_1] + k_2[A_2]}{k_1[A_1] + k_{-2}[B_2]} \tag{4.2.5}$$

The first hypothesis of Temkin is a continuous distribution function of the following form, to be justified later:

$$\boxed{dS' = a \exp(-\gamma A^0/RT) \, d(A^0/RT)} \tag{4.2.6}$$

where γ is a parameter characterizing the distribution and dS' is the number of sites per cm^2 with a standard affinity for adsorption between A^0 and $A^0 + dA^0$ in the ensemble E_j. The constant a will be determined later by means of the condition (4.2.1).

The second hypothesis of Temkin is a Brønsted-type relation between rate constants k_i and equilibrium constants $K_i = k_i/k_{-i}$:

$$\boxed{k_i = \text{Const} \times K_i^\alpha} \qquad (4.2.7)$$

where α is a so-called transfer coefficient, between zero and unity, frequently in the vicinity of 0.5. Relation (4.2.7) is very general in chemical kinetics, e.g. in acid-base catalysis, or the relations of Polanyi and Semenov (see Boudart, 1968).

The Polanyi relation can be understood qualitatively for a family of exothermic elementary reactions with increasing values of the enthalpy of reaction $|\Delta H|$. To an increase $\Delta|\Delta H|$, corresponds a decrease ΔE_a in the activation barrier (Fig. 4.3) such that:

$$|\Delta E_a| = \alpha \, \Delta |\Delta H|$$

Or more simply:

$$E_a = E^0 + \alpha \, \Delta H \qquad (4.2.8)$$

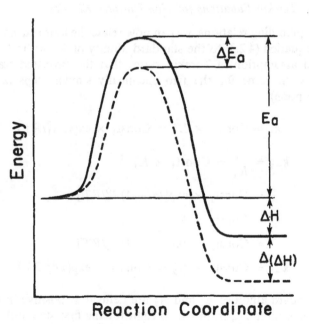

Fig. 4.3 Relation of Polanyi
E_a: activation barrier
ΔH: enthalpy of reaction

where ΔH is negative (exothermic reaction) and $0 < \alpha < 1$ and E^0 is another constant for the particular family of reactions.

The Brønsted relation follows from that of Polanyi:

$$\frac{E_a}{RT} = \text{Const} + \alpha \frac{\Delta H}{RT}$$

$$\exp\left(-\frac{E_a}{RT}\right) = \text{Const}' \times \exp\left(\frac{-\alpha \Delta H}{RT}\right) \qquad (4.2.9)$$

If we assume that the entropy of activation and the entropy of reaction do not change from one reaction to another among the family considered, eq. (4.2.9) becomes:

$$k = \text{Const} \times K^\alpha \qquad (4.2.9a)$$

Similar relations are also used in physical organic chemistry under the name of linear free-energy relationships.

4.22   The Temkin Equations for Non-Uniform Kinetics

From the preceding relations, we can now relate the four rate constants of the rate equation (4.2.3) to the standard affinity of adsorption.

A third assumption of Temkin states that the Brønsted transfer coefficient is the same for the two chemically similar steps (adsorption and desorption):

$$k_1 = \text{Const}_1 \times K_1^\alpha = \text{Const}_1 \times \exp(\alpha A_1^0/RT) \qquad (4.2.10)$$

$$k_{-1} = \frac{k_1}{K_1} = \text{Const}_1 \times K_1^{\alpha-1}$$

$$\dot{=} \text{Const}_1 \times \exp[(\alpha - 1)A_1^0/RT] \qquad (4.2.11)$$

$$k_2 = \text{Const}_2 \times K_2^{\alpha-1}$$

$$= \text{Const}_2 \times \exp[(\alpha - 1)A_2^0/RT] \qquad (4.2.12)$$

$$k_{-2} = \text{Const}_2 \times K_2^\alpha = \text{Const}_2 \times \exp(\alpha A_2^0/RT) \qquad (4.2.13)$$

The standard affinities A_1^0 and A_2^0 are defined as positive in the direction of adsorption, i.e., from left to right for the first step and vice versa for the second step. They are related to each other:

$$A_1^0 - A_2^0 = A_{\text{overall}}^0$$

where $A^0_{overall}$ is the standard affinity of the overall reaction which is a constant in the sense that it depends in no way on the nature of the sites. Hence, only one affinity, say A^0_1 needs to be considered in the above equations.

Let us write for short:

$$t = A^0_1/RT \qquad (4.2.14)$$

when t is a dimensionless affinity, the variation of which, through the distribution function (4.2.6) defines the non-uniformity of the surface. It is assumed that t has an upper value t_0 and a lower value t_1 so that the width of the distribution is:

$$f = t_0 - t_1 \qquad (4.2.15)$$

The rate constants (4.2.10) to (4.2.13) can then be written in a form that leads to an easy integration of eq. (4.2.2).

Thus:

$$k_1 = \text{Const}_1 \times \exp \alpha t$$

But when $t = t_0$, $k_1 = k^0_1$ so that:

$$\text{Const}_1 = k^0_1 \exp(-\alpha t_0)$$

or

$$k_1 = k^0_1 \exp \alpha(t - t_0) \qquad (4.2.16)$$

Similarly,

$$k_{-1} = k^0_{-1} \exp[(\alpha - 1)(t - t_0)] \qquad (4.2.17)$$

$$k_{-2} = \text{Const}_2 \times \exp[\alpha(A^0_1 - A^0_t)/RT]$$

$$= \text{Const}'_2 \times \exp \alpha t$$

Since when $t = t_0$, $\text{Const}'_2 = k_{-2} \exp(-\alpha t_0)$, we have

$$k_{-2} = k^0_{-2} \exp \alpha(t - t_0) \qquad (4.2.18)$$

and finally:

$$k_2 = k^0_2 \exp[(\alpha - 1)(t - t_0)] \qquad (4.2.19)$$

In the expressions (4.2.16) to (4.2.19) for the rate constants, the pre-exponential factors represent the rate constants for the maximum value of the affinity. The constant a of eq. (4.2.6) can now be evaluated:

$$\int dS' = \int_{t_1}^{t_0} a \exp(-\gamma t)\, dt = [L] \tag{4.2.20}$$

Hence:

$$\frac{[L]}{a} = \left[-\frac{1}{\gamma} \exp(-\gamma t) \right]_{t_1}^{t_0}$$

and

$$a = \frac{[L]\gamma \exp(\gamma t_0)}{[\exp(\gamma f)] - 1} \tag{4.2.21}$$

Replacing a by its value above, in eq. (4.2.6), we get:

$$dS = dS'/[L]$$

$$dS = \frac{\gamma \exp(\gamma t_0)}{[\exp(\gamma f)] - 1} \times \exp(-\gamma t)\, dt \tag{4.2.22}$$

where dS is the fraction of sites with affinity between t and $t + dt$.

To integrate eq. (4.2.2), let us introduce the auxiliary variable (eq. 4.2.5) instead of t:

$$u = u_0 \exp(t_0 - t) \tag{4.2.23}$$

with

$$u_0 = \frac{k^0_{-1}[B_1] + k^0_2[A_2]}{k^0_1[A_1] + k^0_{-2}[B_2]} \tag{4.2.24}$$

Clearly, u_0 is the value of u for $t = t_0$. The rate on the collection of sites can be written as:

$$v = \int v_{t,j}\, dS'_j$$

$$= \int \frac{k_1 k_2 [A_1][A_2] - k_{-1} k_{-2}[B_1][B_2]}{k_1[A_1] + k_{-1}[B_1] + k_2[A_2] + k_{-2}[B_2]}\, dS'_j \tag{4.2.25}$$

Let us now introduce the expression for the k_i and dS'_j, regroup the terms and simplify. We get:

$$v = \int_{t_1}^{t_0} \frac{(k_1^0 k_2^0 [A_1][A_2] - k_{-1}^0 k_{-2}^0 [B_1][B_2]) \exp \alpha(t - t_0)}{(k_{-1}^0 [B_1] + k_2^0 [A_2]) + (k_1^0 [A_1] + k_{-2}^0 [B_2]) \exp(t - t_0)}$$

$$\times \frac{[L]\gamma}{[\exp(\gamma f)] - 1} \times \exp - \gamma(t - t_0) \, dt \qquad (4.2.26)$$

Let us write further:

$$m = \alpha - \gamma$$

Since $dt = -du/u$:

$$v = - \int_{u_1}^{u_0} \frac{[L]\gamma}{[\exp(\gamma f)] - 1}$$

$$\times \frac{(k_1^0 k_2^0 [A_1][A_2] - k_{-1}^0 k_{-2}^0 [B_1][B_2]) u_0^m}{(k_{-1}^0 [B_1] + k_2^0 [A_2]) u + (k_1^0 [A_1] + k_{-2}^0 [B_2]) u_0} u^{-m} \, du \qquad (4.2.27)$$

where u_1 is the value of u for $t = t_1$. Replacing now the value of u_0 by its expression (4.2.44) gives:

$$v = - \frac{[L]\gamma}{[\exp(\gamma f)] - 1} \frac{k_1^0 k_2^0 [A_1][A_2] - k_{-1}^0 k_{-2}^0 [B_1][B_2]}{(k_1^0 [A_1] + k_{-2}^0 [B_2])^m (k_{-1}^0 [B_1] + k_2^0 [A_2])^{1-m}}$$

$$\times \int_{u_1}^{u_0} \frac{u^{-m} \, du}{1 + u} \qquad (4.2.28)$$

To evaluate the integral in (4.2.28), let us now use the fourth hypothesis of Temkin concerning the limits u_0 and u_1. It is assumed that for the upper bound value of t, all sites are occupied. This means:

$$u_0 = \frac{[S_1]}{[S_2]} \longrightarrow 0$$

On the other hand, for the lower bound value of t, assume that all sites are empty:

$$u_1 \longrightarrow \infty$$

From tables of definite integrals, we obtain:

$$\int_0^\infty \frac{u^{-m}}{1+u}\,du \simeq \frac{\pi}{\sin(\pi m)} \tag{4.2.29}$$

Hence the final expression of the rate on a non-uniform surface following the formalism of Temkin becomes:

$$v_t = \frac{v}{[L]} = \tau\,\frac{k_1^0 k_2^0 [A_1][A_2] - k_{-1}^0 k_{-2}^0 [B_1][B_2]}{(k_1^0[A_1] + k_{-2}^0[B_2])^m (k_{-1}^0[B_1] + k_2^0[A_2])^{1-m}} \tag{4.2.30}$$

$$\tau = \frac{\pi}{\sin(\pi m)}\,\frac{\gamma}{[\exp(\gamma f)] - 1} \tag{4.2.31}$$

$$m = \alpha - \gamma$$

Note the similarity between expressions (4.2.30) and (4.2.3), the latter for a uniform surface: the numerators are identical, while the denominators contain the same terms, although grouped in different ways. Let us show an application of the Temkin formalism, the first one historically.

4.23 Ammonia Synthesis on an Iron Catalyst

The famous rate equation of Temkin and Pyzhev (1940) was obtained for catalytic ammonia synthesis, corresponding to the overall equation:

$$N_2 + 3H_2 = 2NH_3$$

The already considered two-step mechanism without dissociation of di-nitrogen can be written as:

$$N_2 + * \underset{k_{-1}}{\overset{k_1}{\rightleftarrows}} N_2* \tag{1}$$

$$N_2* + 3H_2 \overset{K_2}{\rightleftharpoons} * + 2NH_3 \tag{2}$$

with $K_2 = k_{-2}/k_2$, N_2* as the most abundant reaction intermediate, and step (1) the rate-determining step. Because of the latter assumption, we can write:

$$k_2^0, k_{-2}^0 \gg k_1^0, k_{-1}^0$$

Applying eq. (4.2.30), with:

$$[A_1] \equiv [N_2], \quad [A_2] \equiv [H_2] \quad \text{and} \quad [B_2] \equiv [NH_3] \quad \text{and} \quad [B_1] \equiv 1$$

we get:

$$v_t = \tau \frac{k_1^0 k_2^0 [N_2][H_2]^3 - k_{-1}^0 k_{-2}^0 [NH_3]^2}{(k_1^0[N_2] + k_{-2}^0[NH_3]^2)^m (k_2^0[H_2]^3 + k_{-1}^0)^{1-m}}$$

With the above inequalities, we obtain:

$$v_t = \tau \left\{ [N_2] k_1^0 \left[\frac{k_2^0}{k_{-2}^0} \frac{[H_2]^3}{[NH_3]^2} \right]^m - k_{-1}^0 \left[\frac{k_{-2}^0}{k_2^0} \frac{[NH_3]^2}{[H_2]^3} \right]^{1-m} \right\}$$

or:

$$v_t = \vec{k}[N_2] \left[\frac{[H_2]^3}{[NH_3]^2} \right]^m - \overleftarrow{k} \left[\frac{[NH_3]^2}{[H_2]^3} \right]^{1-m} \qquad (4.2.32)$$

with:

$$\vec{k} = \tau k_1^0 (K_2^0)^{-m} \qquad (4.2.33)$$

$$\overleftarrow{k} = \tau k_{-1}^0 (K_2^0)^{1-m} \qquad (4.2.34)$$

Thus the overall rate constants can be expressed in terms of those valid for the sites with the highest affinity. Note that the second step is not elementary, but it appears in the rate equation (4.2.32) only through its equilibrium constant K_2^0 and not through its rate constants k_2^0 and k_{-2}^0.

Equation (4.2.32) was first obtained by Temkin and Pyzhev (1940) and is still used, in that form or in a very similar form, in the design of ammonia converters built all over the world.

An attempt was made by Ozaki et al. (1960) to verify the mechanism on which the Temkin-Pyzhev equation is based by comparing on the same catalyst the rates of the two reactions:

$$N_2 + 3H_2 \xrightarrow{K_H} 2NH_3 \qquad (4.2.35)$$

$$N_2 + 3D_2 \xrightarrow{K_D} 2ND_3 \qquad (4.2.36)$$

According to (4.3.32), the ratios of overall rate constants in the forward direction of synthesis with H_2 or D_2 respectively should be:

$$\frac{\vec{k}_H}{\vec{k}_D} = \frac{k_{1,H}^0}{k_{1,D}^0} \left(\frac{K_{2,H}^0}{K_{2,D}^0} \right)^{-m} \qquad (4.2.37)$$

But since hydrogen does not participate in the rds, we should have:

$$k^0_{1,H} = k^0_{1,D} \qquad (4.2.38)$$

Moreover, the ratio $K^0_{2,H}/K^0_{2,D}$ corresponds to the equilibrium constant of the *difference* between the two equilibria:

$$N_2* + 3H_2 \rightleftharpoons 2NH_3 + *$$
$$N_2* + 3D_2 \rightleftharpoons 2ND_3 + *$$

which does not contain $[N_2*]$ or $[*]$:

$$\frac{K^0_{2,D}}{K^0_{2,H}} = \frac{[NH_3]^2[D_2]^3}{[ND_3]^2[H_2]^3} \qquad (4.2.39)$$

The ratio on the left-hand side is simply that of the overall equilibrium constants. Thus:

$$\left(\frac{K^0_{2,H}}{K^0_{2,D}}\right)^{-m} = \left(\frac{K_H}{K_D}\right)^m \qquad (4.2.40)$$

Hence the equation of Temkin-Pyzhev predicts that:

$$\frac{\vec{k}_D}{\vec{k}_H} = \left(\frac{K_H}{K_D}\right)^m \qquad (4.2.41)$$

a result that can be verified readily. This has been done by Shapatina et al. (1971) on an iron catalyst for which it had been already found that $m = 0.5$. Thus m was not an adjusted parameter, and the calculated or experimental values compared in Table 4.1 were obtained without a

TABLE 4.1   Check of the mechanism of Temkin and Pyzhev for ammonia synthesis

Temperature (K)	$\vec{k}_D/\vec{k}_H$ calculated	$\vec{k}_D/\vec{k}_H$ experimental
673	2.87	· 3.13
723	2.60	2.89
753	2.52	2.61

single adjustable parameter. The agreement between theory and experiment is exceptionally good, in view of the total absence of approximations necessary for the calculation.

Consider now the similarity between rate expressions obtained for the same mechanism on uniform or non-uniform surfaces.

For the mechanism as treated above:

$$* \quad + N_2 \quad \xrightarrow{k} N_2* \tag{1}$$

$$N_2* + 3H_2 \overset{K}{=\!\!\ominus\!\!=} * + 2NH_3 \tag{2}$$

the rate equation for a uniform surface is eq. (3.2.86):

$$v = \frac{k[N_2]}{1 + (K[NH_3]^2/[H_2]^3)}$$

But this can be approximated by:

$$v = k'[N_2] \left\{ \frac{[H_2]^3}{[NH_3]^2} \right\}^n \tag{4.2.42}$$

with $0 < n < 1$.

As before, the equation of Temkin gives:

$$v = [L] \frac{\tau k_1^0 k_2^0 [N_2][H_2]^3}{(k_1^0[N_2] + k_{-2}^0[NH_3]^2)^m (k_2^0[H_2]^3)^{1-m}}$$

with $k_1^0[N_2] \ll k_{-2}^{0'}[NH_3]^2$, we have, as before

$$v = \tau[L] k_1^0 \left(\frac{k_2^0}{k_{-2}^0} \right)^m [N_2] \left(\frac{[H_2]^3}{[NH_3]^2} \right)^m$$

or:

$$v = k[N_2] \left(\frac{[H_2]^3}{[NH_3]^2} \right)^m \tag{4.2.43}$$

But this is exactly the result of the approximate expression (4.2.42). Thus, it would be very hard to distinguish between the *exact* expression (3.2.86) for a uniform surface and the *exact* expression (4.2.43) for a non-uniform surface. What will be noted is that is not necessary to invoke the uncomfortable approximation involved in the derivation of (4.2.42), to obtain power rate laws as exemplified by the equation of Temkin and Pyzhev for ammonia synthesis on those representing the data for ammonia or stibine decomposition (§3.2.2b) or the isomerization of pentane (§3.2.2c).

4.3 PHYSICAL MEANING OF THE DISTRIBUTION FUNCTION

Let us now justify a posteriori the formalism of Temkin, as founded on a distribution function, by showing that the latter leads to well-known expressions for adsorption isotherms and rate laws for adsorption.

4.31 Adsorption Isotherms

For the case $\gamma \neq 0$, we obtain Freundlich isotherm. Indeed, the isotherm corresponds to the equilibrium between molecule A and the surface:

$$S_1 \quad + A \underset{k_{-1}}{\overset{k_1}{\rightleftharpoons}} \quad S_2 \tag{1}$$

$$\textit{free site} \qquad\qquad \textit{occupied}$$
$$\textit{site}$$

The fraction of the surface which is occupied is:

$$\theta = \int_{state\ 1}^{state\ 0} \frac{[S_2]}{[S_1] + [S_2]}\, dS \tag{4.3.1}$$

where $[S_2]/([S_1] + [S_2])$ is the fraction of sites of an ensemble E_j and dS the element of the fraction of corresponding sites (eq. 4.2.22). From eq. (4.2.23), we recall that:

$$u = \frac{[S_1]}{[S_2]} = \frac{u_0}{\exp(t - t_0)} = \frac{k_{-1}}{k_1[A]} \tag{4.3.2}$$

or for $t = t_0$:

$$u = u_0 = \frac{k^0_{-1}}{k^0_1[A]} = (K^0_1[A])^{-1} \tag{4.3.3}$$

The fraction of occupied sites in a given ensemble is:

$$\frac{[S_2]}{[S_1] + [S_2]} = \frac{1}{1 + u} = \frac{K_1^0[A]\exp(t - t_0)}{1 + K_1^0[A]\exp(t - t_0)} \qquad (4.3.4)$$

Replacing dS by its expression (4.2.22), we get:

$$\theta = -\int_{u_1}^{u_0} \frac{K_1^0[A]\exp(t - t_0)}{1 + K_1^0[A]\exp(t - t_0)} \frac{\gamma \exp \gamma(t_0 - t)}{[\exp(\gamma f)] - 1} \frac{du}{u} \qquad (4.3.5)$$

where $u_1 = u_0 \exp(t_0 - t_1) = u_0 \exp(f)$. From (4.3.2) and (4.3.3), with:

$$u^{\gamma - 1} = u_0^{\gamma - 1}\exp[(\gamma - 1)(t_0 - t)]$$

we get:

$$\theta = -\frac{\gamma K_1^0[A]}{[\exp(\gamma f)] - 1} \int_{u_1}^{u_0} \frac{u^{\gamma - 1}}{u_0^{\gamma - 1}(1 + u)}\,du \qquad (4.3.6)$$

Since also:

$$u_0^{\gamma - 1} = (K_1^0[A])^{-(\gamma - 1)}$$

we obtain finally:

$$\theta = -\frac{\gamma(K_1^0)^{\gamma}[A]^{\gamma}}{[\exp(\gamma f)] - 1} \int_{u_1}^{u_0} \frac{u^{\gamma - 1}}{1 + u}\,du \qquad (4.3.7)$$

Remembering the third assumption of Temkin, according to which $u_1 \to \infty$ and $u_0 \to 0$, the integral in (4.3.7) is again equal to $\pi/\sin(\gamma\pi)$. Hence:

$$\boxed{\theta = \frac{\gamma(K_1^0)^{\gamma}}{[\exp(\gamma f)] - 1}\frac{\pi}{\sin \gamma\pi}[A]^{\gamma} = \text{Const} \times [A]^{\gamma}} \qquad (4.3.8)$$

This is the form of the empirical equation of Freundlich, used extensively in physisorption and chemisorption. For the latter, an impressive example is the adsorption of hydrogen on a tungsten powder between 10^{-5} torr and 1 atm, from 373 to 1023 K (Frankenburg, 1944). Thus the Freundlich isotherm corresponds to an exponential distribution (Roginskii, 1948).

On the other hand, when γ tends to zero, another isotherm is obtained, that of Frumkin-Temkin. To obtain θ we start from eq. (4.3.6) and note that $\lim_{\gamma \to 0} (\gamma/[\exp(\gamma f)] - 1) = 1/f$. Then:

$$\theta = \frac{1}{f} \int_{u_0}^{u_1} \frac{du}{(1 + u)u} \tag{4.3.9}$$

or

$$\theta = 1 - \frac{1}{f} \ln \frac{1 + u_0 \exp(f)}{1 + u_0} \tag{4.3.10}$$

since $u_0 = (K_1^0[A])^{-1}$:

$$\theta = 1 - \frac{1}{f} \ln \frac{K_1^0[A] + \exp(f)}{1 + K_1^0[A]} \tag{4.3.11}$$

If the distribution is wide enough, so that $\exp(f) \gg K_1^0[A]$:

$$\theta = \frac{1}{f} \ln(1 + K_1^0[A]) \tag{4.3.12}$$

If moreover A is strongly adsorbed and the pressure is not too small, $K_1^0[A] \gg 1$ and θ becomes:

$$\boxed{\theta = (1/f) \ln K_1^0[A]} \tag{4.3.13}$$

This is the isotherm of Frumkin-Temkin. Examples are the adsorption of nitrogen on iron (Scholten et al., 1959) or molybdenum (Boudart et al., 1981).

4.32 Meaning of the Parameter γ of the Distribution Function

Consider $S'(t)$, the total number of sites per cm^2 with an affinity of adsorption without dimension $t = A^0/RT$, between t and t_0:

$$S'(t) = \int_t^{t_0} dS'$$

$$= \frac{[L]}{[\exp(\gamma f)] - 1} \{[\exp \gamma(t_0 - t)] - 1\} \tag{4.3.14}$$

Thus, the fraction of sites: $S(t) = S'(t)/[L]$ with an affinity larger than t is:

$$S(t) = \{[\exp \gamma(t_0 - t)] - 1\}/\{[\exp(\gamma f)] - 1\} \qquad (4.3.15)$$

Hence:

$$t_0 - t = \frac{1}{\gamma}\ln\{1 + S\{[\exp(\gamma f)] - 1\}\} \qquad (4.3.16)$$

and if $\gamma \to 0$

$$t_0 - t = fS \qquad (4.3.17)$$

Alternatively:

$$A^0 = A_0^0 - CS \qquad (4.3.18)$$

with

$$C = fRT \qquad (4.3.19)$$

Thus, the standard affinity of adsorption decreases linearly with S. It can also be shown that for $\varepsilon < \theta < 1 - \varepsilon$ with $\varepsilon \ll 1$, we can write:

$$A^0 = A_0^0 - C\theta \qquad (4.3.20)$$

as $S = \theta$ within an excellent approximation except for very low or very high values of θ (Khammouma, 1972). The difference between the cases $\gamma = 0$ and $\gamma \neq 0$ is best manifested by the difference in the variation of the *affinity of adsorption* with S or θ. Although the latter is not usually specified, there are many studies of the variation of the *heat of adsorption* with coverage, as obtained from calorimetry or from adsorption isotherms (isosteric heats). The variation of heats of adsorption with coverage is due to two factors: first, the existence at the surface of sites with different binding energies (a priori or biographical heterogeneity); second, the interaction between adsorbed species (sometimes called induced heterogeneity). The first factor dominates at low values of coverage, and the second at high values.

As shown by Sips (1950), the analytical form of adsorption isotherms (Freundlich, Frumkin-Temkin, or others) depends on the form of the distribution function of adsorption energies, as embodied in the formalism of Temkin. Since the physical meaning of these distribution functions rests on a priori heterogeneity or interaction between adsorbed species, the formalism actually accounts for these two physical effects responsible for the non-ideality of adsorbed phases. The latter can also be described by means of a thermodynamic activity, replacing surface concentrations. This will be seen later (§4.54) where results very similar to those of the Temkin formalism will be obtained.

4.33 Rate of Adsorption: Equation of Elovich

This equation can be obtained by considering a two-step sequence:

$$A_1 + S_1 \overset{A}{\longrightarrow} S_2 \tag{1}$$

$$\underset{\text{free}}{} \underset{\text{site}}{} \underset{\text{occupied}}{} \underset{\text{site}}{}$$

$$S_2 \rightleftharpoons B_2 + S_1 \tag{2}$$

where step (1) is rate determining and step (2) is equilibrated. We will treat the case of $\gamma \to 0$ corresponding to the Frumkin-Temkin isotherm:

$$\theta = (1/f)\ln K_2^0[B_2] \tag{4.3.21}$$

The rate expression is:

$$v = v_1 = \int [L]\frac{k_1 k_2[A_1]}{k_2 + k_{-2}[B_2]} ds \tag{4.3.22}$$

With previous results, this becomes, according to the Temkin equation:

$$v_1 = [L]\frac{\gamma}{[\exp(\gamma f)] - 1}\frac{\pi}{\sin(\pi m)} k_1^0[A_1](K_2^0[B_2])^{-m} \tag{4.3.23}$$

As $\gamma \to 0$ and with (4.3.21) we get:

$$\boxed{v = \frac{[L]}{f}\frac{\pi}{\sin(\pi\alpha)} k_1^0[A_1]\exp(-g\theta)} \tag{4.3.24}$$

where $g = \alpha f$.

Thus the rate of adsorption decreases exponentially as θ increases. This is the result of Elovich (Low, 1960; Aharoni and Tompkins, 1970) also associated with the names of Zeldovich and Roginskii. The equation has been used extensively to describe rates of chemisorption. A similar equation, eq. (2.5.15), had been proposed earlier by Langmuir to account for the exponential decrease in rates of desorption as θ decreases. Examples of the Elovich and Langmuir equations are the adsorption of nitrogen on iron (Scholten et al., 1959) and the desorption of nitrogen from molybdenum (Boudart et al., 1981).

Thus the formalism of Temkin for ammonia synthesis on iron is supported by overall kinetic data, by the form of adsorption isotherms of nitrogen, and by the rate of adsorption of nitrogen. But more importantly perhaps, the formalism leads to some general useful results that will now be discussed.

4.4 CONSEQUENCES OF TEMKIN'S FORMALISM

Temkin's equation was derived for a two-step sequence or for any reaction that can be condensed to such a sequence. But for such a condensation to be possible, certain simplifications must be made. Is it legitimate to assume a rate-determining step in the case of a non-uniform surface?

4.41 The Concept of Rate-Determining Step is Preserved on a Non-Uniform Surface

This concept, as seen in Chapter 3, was introduced to recognize the possibility that the affinity of a step is different from zero, but is equal to zero (or almost equal to zero) for all the other steps that are equilibrated or quasi-equilibrated. But the treatment of non-uniform surfaces introduces a parameter $t = A^0/RT$ that varies from a maximum value t_0 to a minimum value t_1 over a range $f = t_0 - t_1$.

By definition the adsorption equilibrium constant K depends exponentially on t: $K = \exp(t)$. Following Brønsted, we also admit that the rate constant for adsorption, k, can be written as $k = \text{Const.} \times K^\alpha$. In ammonia synthesis, this means that the rate constant for adsorption of dinitrogen is large for sites with $t = t_0$ but small for sites with $t = t_1$.

It would be incorrect to think that the adsorption step cannot be rate-determining on sites with a high value of the affinity for adsorption. According to this reasoning, the rate-determining step could be a subsequent one involving the hydrogenation of tenaciously held nitrogen, or even the desorption of strongly bound ammonia. But this is only half of the story. Indeed the rate of adsorption of dinitrogen is:

$$v_{\text{ads}} = k_{\text{ads}}[N_2][*]$$

when $t = t_0$, k_{ads} is indeed large, but $[*]$ is quite small as most sites are occupied. Conversely, on weak sites ($t = t_1$), k_{ads} is small, but many such sites are free, so that $[*]$ is large.

Thus we expect a compensation effect which can be quantified readily by expressing rate constants as functions of t in the values of the forward rates of the two steps:

$$S_1 + A_1 \rightleftharpoons S_2 + B_1$$

$$A_2 + S_2 \rightleftharpoons S_1 + B_2$$

Clearly:

$$\frac{\vec{v}_1}{\vec{v}_2} = \frac{k_1[A_1][S_1]}{k_2[A_2][S_2]} \tag{4.4.1}$$

We now replace in (4.4.1), k_1 and k_2 by the relations (4.2.16) and (4.2.19), and $[S_1]/[S_2]$ by $u = u_0 \exp(t - t_0)$ following (4.2.23):

$$\frac{\vec{v}_1}{\vec{v}_2} = \frac{k_1^0 \exp \alpha(t - t_0)[A_1]}{k_2^0 \exp[(\alpha - 1)(t - t_0)][A_2]} u_0 \exp - (t - t_0)$$

Hence:

$$\frac{\vec{v}_1}{\vec{v}_2} = u_0 \frac{k_1^0 [A_1]}{k_2^0 [A_2]} \tag{4.4.2}$$

Consequently this ratio does not depend on t but only on t_0, the highest value of t. It is concluded that if $\vec{v}_1 \ll \vec{v}_2$ for sites with $t = t_0$, then $\vec{v}_1 \ll \vec{v}_2$ for all sites. In other words, if step (1) is rate determining on the sites with the highest value of the affinity, it is rate determining on all other sites (Madix, 1968). The concept of a rate-determining step remains valid within the formalism of Temkin.

4.42 The Search for the Best Site or the Best Catalyst

On a non-uniform surface, which are sites with the highest turnover rate? Consider once more a two-step reaction. The turnover rate on a given site among a collection of non-uniform sites, or on a given catalyst among a collection of catalysts, is given by:

$$v_t = \frac{k_1 k_2 [A_1][A_2] - k_{-1} k_{-2} [B_1][B_2]}{k_1[A_1] + k_{-1}[B_1] + k_2[A_2] + k_{-2}[B_2]}$$

To simplify matters, let us write $\alpha = 1/2$ for the Brønsted exponent. Then substitution of the values of $k_{\pm i}$ with $\alpha = 1/2$ (eqs. 4.2.16 to 4.2.19) yields:

$$v_t = \frac{k_1^0 k_2^0 [A_1][A_2] - k_{-1}^0 k_{-2}^0 [B_1][B_2]}{E_1 \exp \frac{1}{2}(t - t_0) + E_2 \exp \frac{1}{2}(t_0 - t)} \tag{4.4.3}$$

where E_1 and E_2 are defined as:

$$E_1 = k_1^0 [A_1] + k_{-2}^0 [B_2] \tag{4.4.4}$$

and

$$E_2 = k_2^0 [A_2] + k_{-1}^0 [B_1] \tag{4.4.5}$$

For an optimum turnover rate:

$$dv_t/dt = 0$$

Since the numerator of eq. (4.4.3) depends only on t_0 and not on t, all we must write is:

$$dD/dt = 0$$

where D is the denominator of the same equation. It follows that

$$\frac{d}{dt} [E_1 \exp \tfrac{1}{2}(t_{max} - t_0) + E_2 \exp \tfrac{1}{2}(t_0 - t_{max})] = 0$$

where t_{max} is the value of t for $v_{t,max}$. The condition is:

$$E_2/E_1 = \exp(t_{max} - t_0) \tag{4.4.6}$$

Hence

$$\boxed{D_{v_{t,max}} = 2(E_1 E_2)^{\frac{1}{2}}} \tag{4.4.7}$$

But the ratio E_2/E_1 is equal to u_0 (see eq. 4.2.24). Since also, according to eq. (4.2.23):

$$1/\exp(t_0 - t_{max}) = u_0/u_{t_{max}}$$

it follows that

$$\frac{E_2}{E_1} = \frac{1}{\exp(t_0 - t_{\max})} = \frac{u_0}{u_{t_{\max}}} = u_0 \qquad (4.4.8)$$

and

$$u_{t_{\max}} = 1 \qquad (4.4.9)$$

The meaning of this result is simple if we remember that $u = [S_1]/[S_2]$. Thus for the optimum site or catalyst:

$$[S_1] = [S_2]$$

and the fractional surface coverage is equal to 1/2. Since surface coverage is highest on sites with the largest affinity, and vice versa, the conclusion is that the optimum site or catalyst is one with an intermediate value of the affinity. As will be seen below, it corresponds to the center of the distribution for $t = f/2$ where $f = t_0 - t_1$.

The above conclusion, based on a simple model, must be considered in the light of the experimental result (Chapter 3), according to which an iron catalyst for ammonia synthesis is about half-covered with nitrogen during the steady-state reaction. Another way to look at these results is to consider them as expressions of the principle of Sabatier, who considered the optimum catalyst the one capable of making an intermediate compound of sufficient, but not too high stability.

In practice, to optimize a catalyst, what must be done is to change the value of the affinity t. This can be done with metallic catalysts in three different ways. First, the *surface structure* can be changed by changing the crystallographic planes or by changing the particle size. In both cases, this consists of modifying the relative surface population of atoms with different coordination numbers. Second, the metal can be modified by *alloy* formation (e.g., Cu into Ni) or addition of a surface impurity (e.g., sulfur, carbon, oxygen, nitrogen). Third, the search for the optimum catalyst may consist in changing the *nature of the metal* by going over the periodic table. For the comparison to be meaningful, it is assumed that the reaction mechanism does not change. If it does, this may be favorable or not, but that is another matter.

By optimum, we mean only optimum turnover rate or activity. We do not discuss the other essential components of an optimum catalyst: selectivity, porosity, mechanical resistance, and stability against sintering or poisons.

Frequently, soon after introduction of a new catalyst, its activity is improved. After a while, further gains in activity become marginal until a plateau is reached. To go beyond the plateau, it seems that a different sequence of chemical steps must be discovered.

4.43 Variation of Activity from One Site (or Catalyst) to the Next

It is quite instructive how v_t changes between the upper and lower values of t, on both sides of its maximum value $v_{t,\,max}$ for values of t equal to $t_{max} \pm t_i$. Thus, replacing t by $(t_{max} - t_i)$ in eq. (4.4.3), we obtain:

$$v_t = \frac{k_1^0 k_2^0 [A_1][A_2] - k_{-1}^0 k_{-2}^0 [B_1][B_2]}{E_1 \exp \tfrac{1}{2}(t_{max} - t_i - t_0) + E_2 \exp \tfrac{1}{2}(t_0 - t_{max} + t_i)} \qquad (4.4.10)$$

Now, eq. (4.4.6) gives:

$$E_1^{\frac{1}{2}} = E_2^{\frac{1}{2}} \exp \tfrac{1}{2}(t_0 - t_{max}) \qquad (4.4.11)$$

The denominator D of the expression for v_t can be written in the form:

$$D_{t_i} = (E_1 E_2)^{\frac{1}{2}} [\exp(-t_i/2) + \exp(t_i/2)] \qquad (4.4.12)$$

But this is also the result obtained if we replace t by $(t_{max} + t_i)$. Thus v_t changes symmetrically with t_i on both sides of $v_{t,\,max}$ (Fig. 4.4).

In particular, $v_{t,\,max}$ lies at the center of $t_0 - t_i = f$ and the values of $v_{t,\,min}$ on both sides are the same. They correspond to values of:

$$t_i = \frac{t_0 - t_1}{2} = f/2$$

The denominator becomes for those values:

$$D_{v_{t,\,min}} = (E_1 E_2)^{\frac{1}{2}} [\exp(f/4) + \exp(-f/4)] \qquad (4.4.13)$$

To fix the ideas, let f be equal to 10, a reasonable value:

$$f = t_0 - t_1 = (A_0^0 - A_1^0)/RT$$

Indeed, at 750 K, this would correspond to a variation ΔA^0 equal to 15 kcal mol^{-1}, a moderate spectrum of values for different sites or catalysts. Then, the term $\exp(-f/4)$ in eq. (4.4.13) can be neglected and:

$$\boxed{D_{v_{t,\,min}} = (E_1 E_2)^{\frac{1}{2}} \exp(f/4)} \qquad (4.4.14)$$

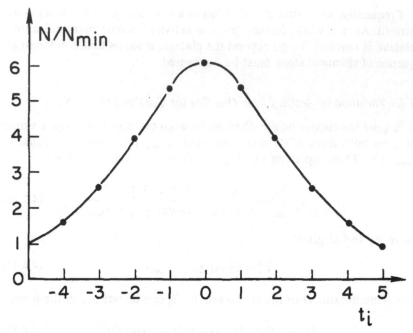

Fig. 4.4 Variation of v_t in the range of values of t defined by $f = 10$: a typical volcano curve

This result must be compared to that for $D_{v_t, max}$ (eq. 4.4.7)

$$D_{v_t, max} = 2(E_1 E_2)^{\frac{1}{2}}$$

The comparison between maximum and minimum turnover rates is:

$$\boxed{\frac{D_{v_t, max}}{D_{v_t, min}} = \frac{\exp(f/4)}{2}} \qquad (4.4.15)$$

The activity of a collection of sites of catalysts changes because of three factors. The first factor is thermodynamic and defined as $K = \exp(t)$ for the higher and lower values of t so that:

$$\boxed{K_{t_0}/K_{t_1} = \exp(t_0 - t_1) = \exp(f)} \qquad (4.4.16)$$

The second factor is kinetic and defined by the Brønsted relations for the rate constants. If $\alpha = 1/2$ for simplicity, the values of the rate constant change by the amount:

$$\boxed{k_{t_0}/k_{t_1} = \exp(f/2)} \tag{4.4.17}$$

The third factor is the one of interest in the catalytic reaction: it is expressed by eq. (4.4.15) and shows that the ratio of maximum and minimum values of the turnover rates is only $1/2 \exp(f/4)$. This is because of a compensation: since k and K change in the same direction, the sites become "better," but there are fewer of them available for reaction.

Another result is interesting. It is the ratio between:

$$v_{t,\max} = \frac{k_1^0 k_2^0 [A_1][A_2] - k_{-1}^0 k_{-2}^0 [B_1][B_2]}{2(E_1 E_2)^{\frac{1}{2}}} \tag{4.4.18}$$

and v_{nu} calculated for a non-uniform distribution with $\gamma = 0$ (see §4.32), corresponding to a surface where there is the same number of sites with each value of t between t_0 and t_1. In both cases, we assume $\alpha = 1/2$. Then from the general equation (4.2.30), since $m = \alpha - \gamma = 1/2$, with:

$$\lim_{\gamma \to 0} \frac{\gamma}{[\exp(\gamma f)] - 1} = \frac{1}{f}$$

we get

$$v_{t,\mathrm{nu},\gamma=0,\alpha=\frac{1}{2}} = \frac{\pi}{f} \frac{k_1^0 k_2^0 [A_1][A_2] - k_{-1}^0 k_{-2}^0 [B_1][B_2]}{(E_1 E_2)^{\frac{1}{2}}} \tag{4.4.19}$$

The ratio we want is thus:

$$v_{t,\max}/v_{t,\mathrm{nu}} = f/2\pi \tag{4.4.20}$$

This result is rather striking: the activity of the best site or catalyst is only $f/2\pi$ times larger than that of a non-uniform catalyst with a collection of sites, each ensemble of the collection having the same number of sites with t between t_0 and t_1 with $t_0 - t_1 = f$ (Kiperman, 1964).

In conclusion, it appears that a non-uniform surface behaves catalytically much more like a uniform surface than suspected from a knowledge of its thermodynamic behavior. Rate equations are similar for a given mechanism on a uniform or non-uniform surface. This result justifies the common practice of neglecting non-uniformity of catalytic surfaces in kinetic studies.

Yet, there are cases where different catalysts exhibit very large differences in activity. How can that be?

4.44 Interpretation of Very Large Differences in Catalytic Activity

Such large differences are not readily accessible to experimental verification for sites on a given surface. But they are found when certain reactions are studied on different catalysts. How can they be explained?

According to Boudart (1976), if the reactant involved in the rate-determining step requires several adjacent surface atoms, i.e., a multiple site (ensemble, cluster), a broad spectrum of activities is expected, the broader the larger the size of the cluster.

This is easily seen with a simple two-step sequence:

$$A + n* \xrightarrow{\ k\ } A.n* \tag{1}$$

$$B* \underset{\ }{\overset{K}{\rightleftharpoons}} * + B \tag{2}$$

The overall reaction is $A = B$. The species $B*$ is the most abundant reaction intermediate, and A requires n adjacent surface atoms in the rate-determining adsorption step. The rate is then:

$$v = k[A][*]\left\{\frac{[*]}{[L]}\frac{z}{2}\right\}^{n-1} \tag{4.4.21}$$

where the last factor to the $(n-1)$th power is the probability of finding $(n-1)$ free atoms that are adjacent to the first one, z being the number of nearest neighbors to a given atom (see §2.23). As ever:

$$[L] = [*] + [B*]$$

with

$$K = \frac{[B*]}{[*][B]}$$

Elimination of $[B*]$ and $[*]$ yields:

$$v = k(z/2)^{n-1}[A][*]^n[L]^{-(n-1)}$$

or

$$v = \frac{k'[A]}{(1 + K[B])^n} \tag{4.4.22}$$

with

$$k' = k(z/2)^{n-1}[L]$$

The Brønsted relation with $\alpha = 0.5$ relates the equilibrium constant to the rate constant:

$$k' = CK^{\frac{1}{2}} \qquad (4.4.23)$$

where C is a constant. Substitution gives:

$$v = \frac{CK^{\frac{1}{2}}[A]}{(1 + K[B])^n} \qquad (4.4.24)$$

Suppose that the affinity of adsorption changes as a result of a change in structure, addition of a modifier or change in the nature of the metal. The maximum rate v_0 will be obtained for an optimum value K_0 of K when:

$$(dv/dK) = 0 \qquad (4.4.25)$$

Hence:

$$K_0 = 1/[B](2n - 1) \qquad (4.4.26)$$

Hence:

$$\frac{v}{v_0} = \left(\frac{K}{K_0}\right)^{\frac{1}{2}}\left\{\frac{1 + K_0[B]}{1 + K[B]}\right\}^n \qquad (4.4.27)$$

Let us write:

$$K/K_0 = \zeta \qquad (4.4.28)$$

Then the ratio of rates becomes with the help of eq. (4.4.26):

$$\frac{v}{v_0} = \zeta^{\frac{1}{2}}\left\{\frac{1 + [1/(2n - 1)]}{1 + [\zeta/(2n - 1)]}\right\}^n \qquad (4.4.29)$$

Values of v_0/v are collected on Table 4.2 for sets of values of ζ and n.

TABLE 4.2 Ratio of rates (eq. 4.4.29) (Boudart, 1976)

Ratio of equilibrium constants for adsorption: $\zeta = K/K_0$	Number of atoms n in the multiple chemisorption site					
	1	2	3	4	5	6
10	1.7	3.3	4.9	6.4	7.8	9.1
10^2	5.0	66.3	5.4×10^2	3.2×10^3	1.5×10^4	6.3×10^4
10^3	15.8	2.0×10^2	1.5×10^5	7.9×10^6	3.3×10^8	1.1×10^{10}
10^4	50.0	6.2×10^4	4.6×10^7	2.4×10^{10}	1.0×10^{13}	3.4×10^{15}
10^5	1.6×10^2	2.0×10^6	1.5×10^{10}	7.7×10^{13}	3.2×10^{17}	1.1×10^{21}
10^6	5.0×10^2	6.2×10^7	4.6×10^{12}	2.4×10^{17}	1.0×10^{22}	3.4×10^{26}

For a single adsorption site ($n = 1$), when K changes one hundredfold on both sides of K_0 (from $\zeta = 10^{-1}$ to $\zeta = 10$), the rate varies by a factor of less than two. The compensation effect discussed in the preceding section is very important. But when the multiplicity n of the site goes up, changes in the rate become important. Thus, when $n = 6$, the same hundredfold change in K on both sides of K_0 modifies the rate by a factor of 10^9. To fix the ideas, it is easy to show, since $K = \exp(A^0/RT)$, that a variation in ζ by 10^n is due to a change in the standard affinity by an amount $2n$ kcal mol^{-1} at a temperature of 435 K, commonly found in practice.

The idea of *multiple sites* has been stressed many times since Langmuir, under the name of *ensembles* proposed by Kobozev (1938) for supported clusters, or *multiplets* favored by Balandin (1929) to emphasize the geometry of sites, or *ensembles* again following Ponec and Sachtler (1972) and Dowden (1972) to denote atoms of the same nature at the surface of an alloy, or *clusters* to define (Muetterties, 1975) metal-metal bonded entities in homogeneous or heterogeneous catalysis.

In the present context, these multiple sites explain large differences in catalytic activity. They will be essential in our discussion (Chapter 6) of structure-sensitive reactions.

4.5 DETERMINATION OF THE THERMODYNAMIC ACTIVITY OF ADSORBED SPECIES: WAGNER'S METHOD

There exists another approach, besides that of Temkin, to account for the non-ideality of surfaces. It consists in replacing the concentration of the important adsorbed species by its thermodynamic activity. The problem is how to measure the latter. How to do this has been shown by Wagner (1970) and his school. The method is based on the measurement of an electromotive force or of an electrical conductivity. The thermodynamic activity of an atom at the surface is the same as that in the bulk, provided that thermodynamic equilibrium between surface and bulk is assured by the proper experimental conditions (thin solid samples and high temperatures). Then, for instance, electrical conductivity of the bulk gives the desired information on the surface. The physical property that is used is first calibrated *at equilibrium* in terms of accessible known values of the activity. Then the physical property measured during the catalytic reaction or its component steps, gives the desired values of the activity on a working catalytic surface.

Wagner's method has been applied to metals, e.g., palladium with dissolved hydrogen, or to oxide semiconductors with non-stoichiometric compositions.

A nice example is the water-gas shift reaction on a thin foil of wüstite (FeO), about 15 μm thick, at high temperatures:

$$CO_2 + H_2 \xrightleftharpoons{K} CO + H_2O \qquad (4.5.1)$$

The reaction has been studied in particular by Stotz (1966) as a clear example of the method of Wagner.

Suppose the reaction proceeds in two steps:

$$(v_1) * \quad + CO_2 \underset{k_{-1}}{\overset{k_1}{\rightleftharpoons}} CO \quad + O* \qquad (4.5.2)$$

$$(v_2) O* + H_2 \quad \underset{k_{-2}}{\overset{k_2}{\rightleftharpoons}} H_2O + * \qquad (4.5.3)$$

where O* is the most abundant reaction intermediate. The chosen temperature is 1173 K to achieve equilibrium between the surface and the bulk of the non-stoichiometric $Fe_{1-x}O$ catalyst, where x is the molar fraction of cationic vacancies.

The plan is to verify the mechanism embodied in the above sequence by measuring the thermodynamic activity of the adsorbed oxygen atoms being transferred between the $H_2 - H_2O$ and $CO - CO_2$ pairs of reactants and products. The idea is to measure by a relaxation method, the rate of transfer of oxygen from CO_2 to the surface (v_1) and from the surface to H_2 (v_2). The rates v_1 and v_2 must be equal at the stationary state of the overall reaction. Hence, the curves representing v_1 and v_2 as a function of the activity of O* must intersect (Fig. 4.6) at a value of the activity $a_{O,s}$ corresponding to the rate v_3 of the overall reaction, as measured separately.

4.51 Principle of Relaxation Measurements

Relaxation methods as used in homogeneous kinetics are used increasingly in heterogeneous kinetics, e.g., in molecular beam relaxation spectroscopy as pioneered by Schwarz and Madix (1968) and discussed in §3.22b. Another example is the isotope jump technique of Tamaru (see Chapter 3).

The principle is shown in Fig. 4.5. A reaction mixture with a given ratio of partial pressures (p_{CO_2}/p_{CO}) flows over the wüstite foil at 1173 K. The electrical conductivity of the latter, σ, reaches a constant value σ_1. At time t_1, the reactant ratio is varied suddenly. The concentration of O* increases, and as it does, σ reaches a new equilibrium value. At the high temperature of the experiment, diffusion within the foil is fast enough so

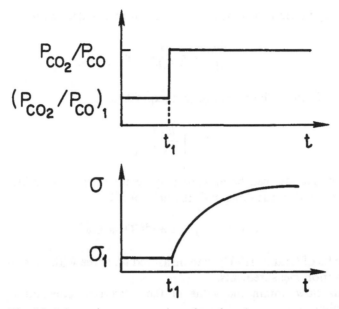

Fig. 4.5 Schematic representation of a relaxation measurement

that the relaxation rate is that of step (4.5.2). Thus the rate of oxygen transfer to the surface can be measured. A similar experiment can be devised with a H_2/H_2O mixture.

4.52 Determination of the Relaxation Constant: Exchange Rates

The mole fraction x is defined by the constant number of atoms of iron (n_{Fe}) and the variable number of oxygen atoms n_O:

$$x = (n_O - n_{Fe})/n_O \qquad (4.5.4)$$

Thus, x changes as follows:

$$dx = \left[-\frac{dn_{Fe}}{dn_O} \right]_{n_{Fe}} = (n_{Fe}/n_O^2)\, dn_O$$

$$\simeq dn_O/n_{Fe} \qquad (4.5.5)$$

as departure from stoichiometry is small ($n_O \simeq n_{Fe}$).

Near equilibrium ($x = x_e$), the relaxation rate is defined by:

$$\frac{dn_O}{dt} = n_{Fe} \frac{dx}{dt} = v_1 \qquad (4.5.6)$$

But, by definition, the exchange rate (eq. 3.2.33) is:

$$v_1^0 = \left[\frac{\partial v_1}{\partial x}\right]_{x=x_e} \qquad (4.5.7)$$

The exchange rate can be measured by means of tracers (§3.21c). In the present case, this was done with the help of ^{14}C:

$$* + {}^{14}CO_2 \rightleftharpoons {}^{14}CO + O*$$

The results of Grabke (1967), who measured v_1^0, agree with those obtained by relaxation measurements.

Let us now obtain the value of the relaxation constant κ_1. Near equilibrium:

$$v_1 = v_1^0(x - x_e) \qquad (4.5.8)$$

Taking eq. (4.5.6) into account, we obtain:

$$\frac{dx}{x - x_e} = -\frac{1}{n_{Fe}}(-v_1^0)\,dt = -\kappa_1\,dt \qquad (4.5.9)$$

where:

$$\kappa_1 = -\frac{v_1^0}{n_{Fe}} \qquad (4.5.10)$$

The last step is to write the relation between the rate constant k_1 and the relaxation constant, so as to obtain v_1. A similar method then yields v_2.

4.53 Determination of Transfer Rates v_1 and v_2

Let us first define the notations. Thermodynamic activities will be denoted by round parentheses ( ), to differentiate them from concentrations, [ ]. All quantities at equilibrium will be characterized by a subscript e. The equilibrium constant of step (4.5.2) is:

$$K_1 = \frac{k_1}{k_{-1}} = \frac{(CO)_e}{(CO_2)_e} \frac{(O*)_e}{(*)_e} \qquad (4.5.11)$$

Thus, the thermodynamic activity of adsorbed oxygen is:

$$(O*)_e = K_1 \frac{(CO_2)_e}{(CO)_e} (*)_e \qquad (4.5.12)$$

We must now choose the standard state. For convenience, it is that of the equilibrated reaction with all activities equal to unity:

$$* + CO_2 \rightleftharpoons CO + O*$$

activity: 1   1   1   1

Consequently:

$$\Delta G_1^0 = 0; K_1 = 1$$

Hence:

$$(O*)_e = \frac{(CO_2)_e}{(CO)_e} (*)_e \qquad (4.5.13)$$

Let us now write:

$$a_0 = (O*)_e/(*)_e$$

This ratio is equal to the ratio of the activities: $(CO_2)_e/(CO)_e$ which may be represented by the ratio of corresponding partial pressures. The latter is then equal to the thermodynamic activity of the adsorbed species, if it is admitted that $(*)_e = 1$.

The equilibrium constant of the overall reaction, K, and of the step (4.5.3), K_2, are related by means of:

$$K = \frac{(CO)_e(H_2O)_e}{(CO_2)_e(H_2)_e} = \frac{K_1}{K_2} = \frac{1}{K_2} \qquad (4.5.14)$$

with:

$$K_2 = \frac{(H_2)_e}{(H_2O)_e} (O*)_e \qquad (4.5.15)$$

Finally:

$$a_0 = (O*)_e = \frac{(CO_2)_e}{(CO)_e} = K \frac{(H_2O)_e}{(H_2)_e} = \frac{1}{K_2} \frac{(H_2O)_e}{(H_2)_e} \qquad (4.5.16)$$

Let us now examine the relaxation rate. Away from equilibrium, we continue to write the reaction rate in terms of concentrations of the ideal gas phase components and of the activity of adsorbed oxygen:

$$v_1 = k_1[CO_2] - k_{-1}[CO](O*)$$

$$v_1 = k_1[CO_2]\left(1 - \frac{1}{K_1} \frac{[CO](O*)}{[CO_2]}\right) \qquad (4.5.17)$$

or, with the chosen standard state:

$$v_1 = k_1[CO_2]\left(1 - \frac{[CO]}{[CO_2]}(O*)\right) \qquad (4.5.18)$$

Note that replacing k_1/k_{-1} by K_1 expressed in terms of the activities, indicates clearly that the rate constants take into account the non-ideality of the surface. Remembering that $(O*) = a_0$, we have:

$$\boxed{v_1 = k_1[CO_2]\left(1 - \frac{a_0}{[CO_2]/[CO]}\right)} \qquad (4.5.19)$$

$$\text{away from equilibrium}$$

This relation represents the rate of transfer of oxygen from CO_2 to the surface, during the relaxation process represented on Fig. 4.5. The partial pressures of CO_2 and CO are constant and imposed, while a_0 as determined by electrical conductivity is the thermodynamic activity of adsorbed oxygen and is obtained by calibration under equilibrium conditions. The time evolution of $(O*)$ corresponds to an instantaneous fictitious virtual atmosphere $[CO_2]_e/[CO]_e$ in equilibrium with the surface of the catalyst and defining a certain non-stoichiometric composition $Fe_{1-x}O$.

Equation (4.5.16) also gives a way to obtain v_2 by means of a reactant mixture $H_2O - H_2$.

Thus the only parameter needed to obtain v_1 is k_1 which is related to the relaxation constant κ_1. What is this relation? Integration between $t = 0$ and t of eq. (4.5.9) gives:

$$\ln \frac{x_{(t)} - x_e}{x_{(t=0)} - x_e} = -\kappa_1 t \qquad (4.5.20)$$

which gives κ_1 from the known relation between x and the electrical conductivity. On the other hand, from the relation (4.5.10) between a_0 and x, we get:

$$\kappa_1 = -\frac{v_1^0}{n_{Fe}} = -\frac{1}{n_{Fe}} \frac{\partial v_1}{\partial a_0} \frac{\partial a_0}{\partial x} \qquad (4.5.21)$$

From eq. (4.5.19), we obtain

$$\frac{\partial v_1}{\partial a_0} = -\frac{k_1[CO_2]}{[CO_2]/[CO]} = -k_1[CO] \qquad (4.5.22)$$

Taking into account the surface area A of the wüstite catalyst, we write:

$$\kappa_1 = -\frac{1}{n_{Fe}} (-A\, k_1[CO]) \frac{\partial a_0}{\partial x}$$

or finally:

$$\boxed{k_1 = \frac{\kappa_1 n_{Fe}}{A[CO]} \frac{\partial x}{\partial a_0}} \qquad (4.5.23)$$

Thus we know k_1 and consequently v_1 as a function of a_0. The results for both v_1 and v_2 may be plotted as a function of a_0 (Fig. 4.6.). The intersection of the two curves gives v_1* and v_2* and the value of a_0^*, hence of the ratio $[CO_2]_e/[CO]_e$ corresponding to the steady-state subscripts. The values of the overall rate v measured separately at the steady state, as well as the corresponding values of $a_{o,s}$ are collected in Table 4.3. The agreement between the two sets of data provides an excellent demonstration of the value of the method of Wagner to elucidate reaction mechanisms of catalytic reactions.

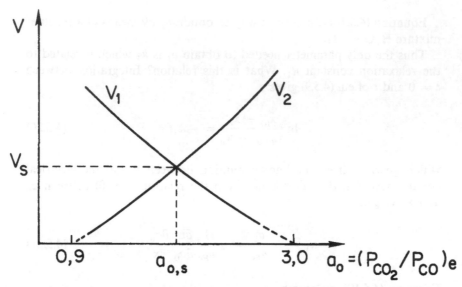

Fig. 4.6 Schematic representation of values of v_1 and v_2 as a function of a_o, for $P_{CO_2}/P_{CO} = 3.0$ and $P_{H_2O}/P_{H_2} = 0.9$. The intersection of the curves corresponds to v_s and $a_{o,s}$ measured directly for the overall water-gas shift reaction at the steady state.

$$v_1: \quad * \quad + CO_2 \rightleftharpoons CO \quad + O*$$
$$v_2: \quad O* \quad + H_2 \rightleftharpoons H_2O + *$$
$$\overline{v: \quad CO_2 + H_2 \quad = \quad CO \quad + H_2O}$$

TABLE 4.3 Comparison between values of a_0 and of rates measured separately and at the steady state on a wüstite sample ($A = 92$ cm^2) at 1173 K

P_{CO} (torr)	P_{H_2O} (torr)	a_0^*	$(a_0)_s$	$v_1 = v_2$ at a_0^* (mol s$^{-1} \times 10^7$)	v at $(a_0)_s$ (mol s$^{-1} \times 10^7$)
543	17.1	1.30	1.33	1.8	2.0
406	17.1	1.15	1.15	1.65	1.7
271	17.1	1.0	1.10	1.10	1.2
136	17.1	0.88	0.89	0.8	1.0
457	71.5	0.91	0.93	1.9	2.2

4.54 Method of Wagner, Formalism of Temkin, Relation of Brønsted

We have seen that the rate of the first step in the overall water-gas shift reaction is given by:

$$v_1 = k_1[CO_2] - k_{-1}[CO](O*)$$

But since: $K_1 = k_1/k_{-1} = 1$:

$$v_1 = k_1([CO_2] - [CO]a_0) \tag{4.5.24}$$

But experimentally, Grabke (1965), and Riecke and Bohnenkamp (1969), found a rate expression of the form:

$$v_1 = k(P_{CO_2}a_0^{-m} - P_{CO}a_0^{1-m}) \tag{4.5.25}$$

or:

$$v_1 = ka_0^{-m}(P_{CO_2} - P_{CO}a_0) \tag{4.5.26}$$

with $m = 0.6$ at 1246 K.

By identification, we find:

$$k_1 = ka_0^{-m}$$

or:

$$k_1 = k[O*]^{-1}a_0^{1-m} \tag{4.5.27}$$

From the relations (4.5.16), we obtain finally:

$$v_1 = k[K^m P_{CO_2}(P_{H_2}/P_{H_2O})^m - (P_{CO}/K^{1-m})(P_{H_2O}/P_{H_2})^{1-m}] \tag{4.5.28}$$

But this is exactly what is obtained from the equation of Temkin (Chapter 3) for a non-uniform surface on the assumption that step (1) is the rate-determining step.

It must be concluded that the method of Wagner and the formalism of Temkin lead to identical results. Following Temkin, the surface is described by a thermodynamic distribution function related to kinetic behavior by the Brønsted relation. This appears to be equivalent to the direct introduction of thermodynamic activity of the adsorbed phase, following Wagner. The problem of Temkin is how to determine the distribution function, and the problem of Wagner is how to determine the thermodynamic activity.

Interesting interpretations at the atomic level of the method of Wagner have been proposed, for instance by Parravano (1970) and by Worrell (1971). But these are not necessary, since Wagner's method is a thermodynamic one.

This concludes our presentation of the kinetic foundations of processes in heterogeneous catalysis. Many examples to follow will illustrate the use of the general treatment presented thus far.

REFERENCES

Aharoni, C. and Tomkins, F. C. 1970. *Advan. Catal. Relat. Subj.* 21:1.

Balandin, A. A. 1929. *Z physik. chem.* B2:289.

Boudart, M. 1968. *Kinetics of Chemical Processes.* Englewood Cliffs, N.J.: Prentice-Hall, p. 167.

Boudart, M. 1976. *Proc. 6th Intl. Congr. Catalysis,* ed. G. C. Bond, P. B. Wells, and F. C. Tompkins, p. 1. London: The Chemical Society, 1977.

Boudart, M., Egawa, S., Oyama, S. T., and Tamaru, K. 1981. *J. Chim. Phys.* 78: 987.

Dowden, D. A. 1972. *Proc. 5th Intl. Congr. Catalysis,* ed. G. C. Bond, P. B. Wells, and F. C. Tompkins, p. 621. New York: American Elsevier.

Frankenburg, W. 1944. *J. Am. Chem. Soc.* 66:1827.

Grabke, H. J. 1965. *Ber. Bunsenges Phys. Chem.* 69:48.

Grabke, H. J. 1967. *Ber. Bunsenges Phys. Chem.* 71:1067.

Khammouma, S. 1972. Ph.D. diss., Stanford University.

Kiperman, S. K. 1964. *Introduction to the Kinetics of Heterogeneous Reactions.* Moscow: Nauka.

Kobozev, N. J. 1938. *Acta physicochim. URSS* 9:805.

Low, M. J. D. 1960. *Chem. Rev.* 60:267.

Madix, R. J. 1968. *Chem. Eng. Sci.* 23:805.

Miller, A. R. 1949. *The Adsorption of Gases on Solids.* Cambridge: Cambridge University Press, p. 19.

Muetterties, E. L. 1975. *Bull. Soc. Chim. Belge* 84:959.

Ozaki, A., Taylor, H., and Boudart, M. 1960. *Proc. Roy. Soc.* (London) A258:47.

Parravano, G. 1970. *Catal. Rev.* 4:53.

Ponec, V. and Sachtler, W. M. H. 1972. *Proc. 5th Intl. Congr. Catalysis,* ed. G. C. Bond, P. B. Wells, and F. C. Tompkins, p. 645. New York: American Elsevier.

Riecke, E. and Bohnenkamp, K. 1969. *Arch. Eisenhutten* 40, 717.

Roberts, J. K. 1935. *Proc. Roy. Soc.* (London) 152:445.

Roginskii, S. Z. 1948. *Adsorption and Catalysis on Non-Uniform Surfaces.* Moscow: N.A.

Schmidt, L. D. 1974. *Catal. Rev. Sci. Eng.* 9:115.

Scholten, J. J. E., Zwietering, P., Konvalinka, J. A., and De Boer, J. H. 1959. *Trans. Faraday Soc.* 55:2166.

Schwarz, J. A. and Madix, R. J. 1968. *J. Catal.* 12:140.

Shapatina, E. I., Kuchaev, V. L., and Temkin, M. I. 1971. *Kinet. Katal.* 12:1476.

Sips, R. 1950. *J. Chem. Phys.* 18:1024.

Stotz, S. 1966. *Ber Bunsenges Phys. Chem.* 70:37.

Temkin, M. I. 1957. *Zhur. Fiz. Khim.* 31:1.

Temkin, M. I. 1965. *Dok. Akad. Nauk. SSSR* 161:160.

Temkin, M. I. and Pyzhev, V. 1940. *Acta Physicochim. URSS* 12:217.

Wagner, C. 1970. *Advan. Catal. Relat. Subj.* 21:323.

Worrell, W. L. 1971. *Advan. High Temp. Chem.* 4:71.

NOTE: Recently, the method of Wagner based on electrochemical measurements has been revived and developed for several catalytic reactions (oxidation, hydrogenation) by H. Saltsburg and co-workers (C. Vayenas and H. Saltsburg, *J. Catal.* 57 [1979]: 296 and H. Saltsburg and M. Mullins, *Annals,* NY Acad. Sci., 415 [1984]: 82).

Chapter 5

STRUCTURE-INSENSITIVE AND STRUCTURE-SENSITIVE REACTIONS ON METALS

5.1 OPERATIONAL DEFINITIONS

5.11 Structure-Sensitive Reactions

Consider a series of catalyst samples, each one consisting of metal particles of size d on an inert support, with d values varying from 1 to 10 nm from sample to sample. As d increases, the relative concentrations of surface atoms C_i and sites B_j change, as i increases from low to high values. This means that the surface *structure* changes (see §2.1). If, then, the turnover frequency for a reaction varies with d as shown on Fig. 5.1, the reaction is said to be structure-sensitive (Boudart, 1969). It must be shown of

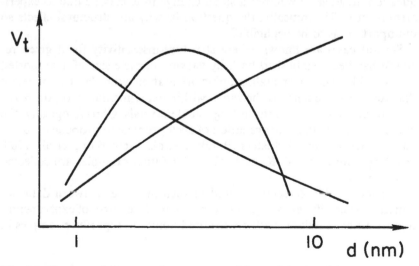

Fig. 5.1 Changes of turnover frequency v_t with particle size d for a structure-sensitive reaction

TABLE 5.1   Schematic change of areal rate with
crystallographic orientation for a structure-sensitive
reaction: ammonia synthesis on iron single crystals

Surface	Areal rate: $n\ mol\ NH_3\ cm^{-2}\ s^{-1}$
(111)	46.00
(100) ordered	2.80
(110) ordered	0.11
(100) disordered	4.50
(110) disordered	0.97

CONDITIONS: Pressure, 2.03 MPa; stoichiometric $N_2 - H_2$ mixture,
$T = 798$ K (Spencer et al., 1982)

course that the variation in v_t is not an artifact, due, for instance, to an effect of support (see Chapter 6). Because of such complication, a safer test of structure sensitivity is one performed on large single crystals exposing well-defined crystallographic planes of varying Miller indices (Table 5.1). In both cases, variations of v_t should be attributed, if possible, to variations in kinetic (E, A) and thermodynamic (ΔS_{ads}^0, ΔH_{ads}^0) parameters in the rate equation. Ultimately, the question is: what is the cause of structure sensitivity?

5.12 Structure-Insensitive Reactions

When v_t does not depend on particle size or crystallographic plane, the reaction is said to be structure-insensitive (Boudart, 1969). Practically, the question arises as to whether a small change in v_t may be due to experimental error. Theoretically, the question is: why are structural effects so unimportant as to be negligible?

Several cases are known where structure-insensitivity for a given reaction has been established both by varying particle size of a supported metal and by varying crystallographic orientation, with the striking result that turnover frequency is the same, within experimental error, on particles about 1 nm in size and on large single crystals. Chronologically, the first example was the hydrogenation of cyclopropane to propane on platinum, either supported clusters (Boudart et al., 1966; Wong et al., 1980) or a large single crystal (Kahn et al., 1974). Other examples are collected in Table 5.2.

The remarkable agreement found in such instances between data obtained on such different samples is a tribute to the control of experimentation in surface catalysis, with an acquired ability to reproduce samples in

TABLE 5.2   Reactions that have been shown to be structure-insensitive, with values of turnover frequency in excellent agreement on large and small crystals

Metal	Reaction	Large single crystals	Supported metallic clusters
Pt	c-C_3H_6 + H_2	Kahn et al. (1974)	Boudart et al. (1966) Wong et al. (1980)
Ni	C_2H_4 + H_2 C_6H_6 + 3 H_2	Dalmai-Imelik and Massardier (1977)	Dalmai-Imelik and Massardier (1977)
{Ni Rh	CO + 3 H_2	Goodman et al. (1980) Somorjai et al. (1980)	Vannice (1976)
Pt	c-C_6H_{10} + H_2	Davis and Somorjai (1980)	Segal et al. (1978)
Pd	CO + $\frac{1}{2}O_2$	Engel and Ertl (1979)	Ladas et al. (1981)

different laboratories and to prepare supported catalysts with metal surfaces meeting the specifications of cleanliness achieved in surface science. Let us now give some details on a reaction shown to be structure-insensitive by both tests, i.e., on small particles and on different crystallographic planes.

5.2 LOW-PRESSURE OXIDATION OF CO ON PALLADIUM: AN EXAMPLE OF A STRUCTURE-INSENSITIVE REACTION

5.21 Study of Engel and Ertl (1978) on a Single Crystal

The kinetic results of Ertl's group have already been described earlier (§3.22). The elementary steps and salient details will be recalled here briefly:

$$CO + * \underset{k_{-1}}{\overset{k_1}{\rightleftharpoons}} CO* \quad (k_1/k_{-1} = K_1) \tag{1}$$

$$O_2 + * \overset{k_2}{\longrightarrow} O*O \tag{2}$$

$$O*O + * \xrightarrow{k_2'} 2O* \qquad (2')$$

$$O* \quad + CO* \xrightarrow{k_3} CO_2 + 2* \qquad (3)$$

The turnover rate v_t was measured at pressures of CO and O_2 in the vicinity of 10^{-4} Pa, between 400 and 700 K. As shown on Fig. 3.4, v_t goes through a pronounced maximum above 500 K. At temperatures below the maximum (region 1), Engel and Ertl conclude that the measured rate is that of O_2 chemisorption (step 2) on a surface largely covered with CO*. The rate equation (3.2.97) is:

$$v = k_2[O_2]/K_1[CO] \qquad (5.2.1)$$

By contrast, above the maximum (region 2), the rate is that of the Langmuir-Hinshelwood step (3), adsorbed CO reacting with O* at the periphery of O* islands with a constant value of [O*]. Since coverage by CO is now small, the rate equation (3.2.99) is now:

$$v = k_3K_1[CO] \qquad (5.2.2)$$

Near the maximum the turnover rate is very high and corresponds to the rate of chemisorption of CO with a sticking probability close to unity.

5.22 Effect of Surface Structure

To find out whether the oxidation of CO on palladium was structure-sensitive or not, Ladas et al. (1981) studied the reaction on metal particles evaporated on a single crystal of α-Al_2O_3 under ultrahigh vacuum conditions. The particle size was measured by transmission electron microscopy, and the number of surface palladium atoms was determined by temperature programmed desorption of CO.

As can be seen on Fig. 5.2, at 443 K, i.e., in region 1, turnover rates are independent of particle size in a critical range between 1.6 and 8 nm, and these are equal to the value reported for the (111) face of a large single crystal. The reaction under study appears to be totally structure-insensitive, a result anticipated by Ertl and Koch (1973), who had seen only very small differences in rates of oxidation of CO over several faces of single crystals of palladium, (100), (110), (111), and (211), as well as on a polycrystalline filament. Yet, as reported by Engel and Ertl (1979), the binding energy of CO on these crystallographic planes of palladium varies by about 7 kcal mole^{-1}, a non-negligible amount.

When the rate is measured at a temperature near the rate maximum (518 K), the agreement between data on Pd (111) and on small particles

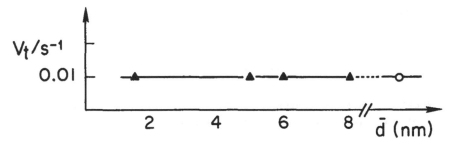

Fig. 5.2 Turnover rate for CO oxidation versus average particle size $\bar{d}$ of palladium particles. The data of Ladas et al. (1981), (▲) are compared to the result reported by Engel and Ertl (1979) on the (111) face of a single crystal of Pd (○). The conditions in both cases are $[O_2]/[CO] = 1.1$; $[CO] = 1.4 \times 10^{-4}$ Pa; $T = 445$ K.

of Pd is again excellent (Fig. 5.3) except for particles smaller than 4 nm. For the latter, the rate is higher than for larger particles. It is higher by about a factor of three for the smallest particles (1.6 nm). It is not clear why the reaction which is so clearly structure-insensitive at 445 K would become structure-sensitive at 518 K for the smallest particles.

A possible explanation (Ladas et al., 1981) may be that surface atoms on clusters are more accessible to gas molecules than surface atoms on flat surfaces. Indeed, following gas kinetic theory, the number of collisions

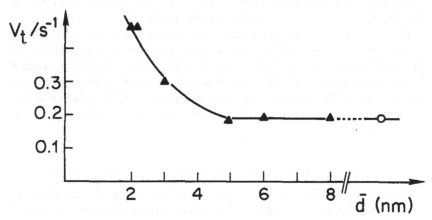

Fig. 5.3 Results of Ladas et al. (1981), (▲) on Pd particles of size $\bar{d}$, near the maximum rate: $T = 518$ K, $[O_2]/[CO] = 1.1$, $[O_2] = 1.2 \times 10^{-4}$ Pa. Comparison with the value of Engel and Ertl (1978) for Pd (111) under the same conditions (○).

of a gas with a number density n cm^{-3} with a surface is, per cm^2 and per second:

$$v_{\text{coll}} = \frac{\bar{v}}{4} n \quad \text{(eq. 1.4.2)}$$

where $\bar{v}$ is the mean molecular velocity. On the other hand, the number of collisions per molecule in the gas phase is:

$$v = \sqrt{2}\pi\sigma^2\bar{v}n \qquad (5.2.3)$$

where $\sqrt{2}\bar{v}$ is now a mean molecular velocity based on reduced mass, and σ is the diameter of the metal atom taken to be equal to that of the gas molecules, the two being considered as identical particles.

Then, per unit surface area of metal atom, the number of collisions per second is given by:

$$\sqrt{2}\bar{v}n \qquad (5.2.4)$$

The ratio of expressions (1.4.2) and (5.2.4) shows that the number of collisions is $4\sqrt{2}$ larger per unit surface area and per unit time for the isolated atom than for a plane surface.

If now we consider a cluster with a high proportion of surface atoms in corner or edge position, we can understand qualitatively why the rate of reaction near the maximum, which is that of adsorption of CO, is higher for clusters than for larger particles. It does not appear necessary to invoke structure-sensitivity to explain the result.

This example of a structure-insensitive reaction is somewhat baffling, as oxygen under the conditions of reaction does not seem to reconstruct the metal surface, as may be the case in other systems. Similarly, there is no carbon deposition which might also erase structural effects. Structure-insensitivity, as sharply pronounced as in the case of CO oxidation at low pressure on palladium, is a difficult phenomenon to understand even qualitatively.

5.23 Thermokinetic Diagram and Improvement of Catalytic Activity

A necessary condition for the complete understanding of a catalytic reaction includes the knowledge of a thermokinetic diagram (see §1.65). For the oxidation of palladium on Pd (111), this diagram is known in great detail from the work of Engel and Ertl (1978). It is shown in Fig. 5.4, which suggests a possible way to improve the catalytic activity by decreasing the binding energy of *CO or *O and, hence, decreasing the

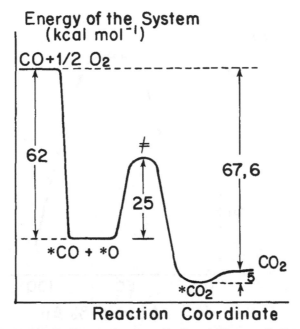

Fig. 5.4 Thermokinetic diagram for the oxidation of CO on palladium (111), after Engel and Ertl (1978)

activation barrier of the Langmuir-Hinshelwood surface step. In fact, the activation barrier for the latter may go down to 14 kcal mol^{-1} if the number density of O* species in the surface islands can be increased (Engel and Ertl, 1978).

The question then arises, is it possible to modify the palladium surface so that the overall adsorption energy of the reaction:

$$2* + CO + \tfrac{1}{2}O_2 \Longrightarrow CO* + O*$$

is decreased, so that the next activation barrier will also decrease? This can be done by formation of an alloy and its possible consequence: a ligand effect (Sachtler and Van der Plank, 1969). However, if the binding energy of reactants with the surface became too small, the dissociation of O_2 might become rate-determining.

This illustration of the principle of Sabatier (see §1.6) is shown in Fig. 5.5, dealing with what appears to be a ligand effect of gold on palladium for the reaction (Lam et al., 1977):

$$H_2 + \tfrac{1}{2}O_2 \Longrightarrow H_2O$$

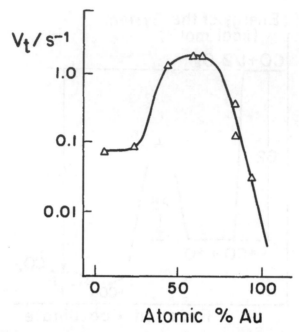

Fig. 5.5 Variation of the turnover frequency for the reaction $H_2 + 1/2\,O_2 = H_2O$ at 263 K, 1.0 kPa H_2, excess O_2, with increasing amount of Au in Pd (Lam et al., 1977)

The reaction was shown to be structure-insensitive in excess oxygen, as the metallic particle size was varied between 1.5 and 4 nm. Since Pd and Au are mutually soluble, a series of alloys could be prepared. They were characterized by means of gold Mössbauer effect spectroscopy, selective chemisorption, and X-ray diffraction (for particles larger than 2.5 nm, in the latter case). It was shown in particular that the surface composition of the alloy particles was the same as the nominal or bulk composition.

It is seen from Fig. 5.5 that the turnover rate first increases as the gold content of the alloy increases, in spite of the fact that gold is about 50 times less active than palladium, justifying its use as an "inert" diluent of the active metal. The rate enhancement with increasing amounts of gold can be understood by a decrease in the binding energy of oxygen on palladium with increasing amounts of gold, as measured by temperature programmed desorption (Weissman-Wenocur and Spicer, 1982). Since the reaction under study is not structure-sensitive, it appears that the observed effect is due to the ligand effect of gold on palladium.

A kinetic study revealed that the reaction rate is proportional to H_2 pressure for alloys containing less than about 60 percent Au. For alloys richer in gold, the kinetics becomes complicated and depends on both pressures of H_2 and O_2. The chemisorption of O_2 on gold-rich alloys becomes influential kinetically.

At any rate, the change of v_t with alloy composition (Fig. 5.5) is yet another illustration of the Sabatier principle, or the "volcano curves," described so many times by Balandin (1969). The rate of reaction first increases with surface binding energy of a key reactant, then reaches a maximum and starts to decline as this binding energy becomes too high.

5.3 AMMONIA SYNTHESIS ON IRON: A STRUCTURE-SENSITIVE REACTION

The results of Ertl (1981) and of Spencer et al. (1982) contribute a great deal to the mechanism of the reaction

$$N_2 + 3H_2 = 2NH_3 \tag{5.3.1}$$

as discussed previously (§3.22b and 4.23) and to the final proof of its structure-sensitivity.

5.31 Overview of Kinetic Results

There are many facts in support of the two-step mechanism with N* as the *mari* (Morikawa and Ozaki, 1971):

$$N_2 + 2* \xrightarrow{\ A\ } 2N* \tag{1}$$

$$N* + 3/2\,H_2 \rightleftharpoons NH_3 + * \tag{2}$$

If the reaction is conducted far away from equilibrium, the dissociative chemisorption of dinitrogen (step 1) is practically irreversible and constitutes the rate-determining step. The overall equilibrium (2) may be treated formally like an elementary step since it carries no kinetic meaning.

Rate expressions found experimentally (eq. 4.2.42) can be obtained from the above mechanism (eqs. 3.2.86 and 4.2.43). It was also shown earlier that, at lower temperatures, the *mari* may contain hydrogen, e.g., as in NH*. Then, the rate equation (eq. 3.2.91) is not very different from the one obtained with N* as *mari*.

More direct kinetic evidence in favor of step (1) being the rate-determining step was obtained from the isotope effect on rates with H_2 and D_2 (§4.23) as well as by determination of $\bar{\sigma}$, the average stoichiometric number of the reaction.

Yet another kinetic and thermodynamic study in favor of the above mechanism is that of Scholten et al. (1959) and Mars et al. (1960). The underlying idea of this work was: if step (1) is the rate-determining step, the rate of adsorption of N_2 must be equal to the rate of synthesis under well-defined conditions, since the rate of adsorption is known to obey the Elovich equation (4.3.24):

$$v_{ads} \propto \exp(-const \times \theta) \qquad (5.3.2)$$

In any comparison with v_{ads}, it is thus necessary to specify the value of the surface coverage θ for which the comparison is made.

An iron catalyst was placed in the bucket of a vacuum microbalance in a stream of either dinitrogen or a stoichiometric mixture of N_2 and H_2. Weight changes are attributed to adsorbed nitrogen atoms, whatever hydrogen there is being neglected because of its small mass. Thus, θ can be monitored at any moment during chemisorption or synthesis. Columns 2 and 3 in Table 5.3 show the values of θ, *at equal rates* of synthesis and chemisorption respectively. The excellent agreement at various temperatures confirms the view that chemisorption of dinitrogen is the rate-determining step.

A third measure of θ confirms this view, θ_{isot} calculated from a separately measured adsorption isotherm following Frumkin-Temkin

$$\theta = (1/f) \ln K_1^0(N_2)_e$$

where, by definition of an equilibrium isotherm, $(N_2)_e$ corresponds to the pressure of dinitrogen at equilibrium with the surface of the catalyst.

However, at the kinetic steady state, H_2 and NH_3 in the gas phase are equilibrated with $N*$, the *mari*, but the latter is not in equilibrium with gaseous N_2. Thus, to calculate θ_{isot}, it is the virtual pressure (see below) and not the equilibrium pressure that must be substituted in the Frumkin-Temkin isotherm. The columns 4 and 5 of Table 5.3 show the

TABLE 5.3 Fraction of iron θ covered with nitrogen

$T(K)$	θ_{syn}	θ_{chem}	θ_{isot}	$P(N_2)_v$, torr
1	2	3	4	5
483	0.44	0.45	0.27	6.7×10^{-4}
522	0.45	0.46	0.40	4.0×10^{-2}
549	0.51	0.53	0.47	6.3×10^{-1}
568	0.54	0.52	0.55	1.5

NOTE: θ_{syn}, θ_{chem}, and θ_{isot} correspond respectively to synthesis, chemisorption, and the adsorption isotherm of Frumkin-Temkin.

corresponding values of the virtual pressure $P(N_2)_v$ and θ_{isot} at various temperatures. Except at the lowest temperature, in spite of the large change in virtual pressure, the value of θ varies but little, is close to those of columns 2 and 3, and is in the vicinity of 1/2. These observations are explained respectively by the logarithmic form of the isotherm, the self-consistency of the assumed mechanism, and the behavior of a non-uniform surface (eq. 4.4.9) with the optimum catalyst corresponding to a value of θ equal to 1/2.

The concept of virtual pressure was introduced by Temkin and Pyzhev (1940) and further developed by Kemball (1966). It is well illustrated by the formation of iron nitride Fe_4N at 673 K. At this temperature, direct reaction between iron and dinitrogen leads to the nitride only if the pressure is higher than 100 MPa, exceeding the decomposition pressure of the nitride. If, however, N_2 is replaced by NH_3, even at only 100 kPa, the nitride is formed, as if a virtual pressure of N_2 exceeding 100 MPa had been used.

The virtual pressure of N_2 can indeed be kept very high, as the equilibrium constant of the reaction of ammonia decomposition:

$$2NH_3 \rightleftharpoons N_2 + 3H_2$$

is equal to ~ 6000 at 673 K, with partial pressures in atm. Thus, if the partial pressures are of the order of 1 atm, the virtual pressure of dinitrogen can be of the order of 100 MPa, and goes up as the partial pressure of H_2 goes down.

In the catalytic synthesis or decomposition of ammonia, adsorbed N* is, by assumption, in equilibrium with gaseous NH_3 and H_2. By contrast, gaseous N_2 is not in equilibrium with N*. The virtual pressure of N_2 can be understood from the following diagrams:

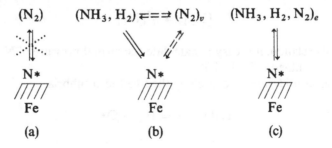

The dashed arrows represent virtual equilibria (case b). Since both systems b and c are equivalent, the equilibrium constant for case c can be used to define the virtual pressure of N_2:

$$K = \frac{[NH_3]_e^2}{[N_2]_e[H_2]_e^3} = \frac{[NH_3]^2}{[N_2]_v[H_2]^3} \qquad (5.3.3)$$

Subscripts e and v refer respectively to concentrations in an equilibrium mixture and in a mixture with a virtual concentration of N_2 that would be in equilibrium with $N*$ and with concentrations of NH_3 and H_2, without subscript, actually present during ammonia synthesis or decomposition. From eq. (5.3.3), $[N_2]_v$ is obtained:

$$[N_2]_v = \frac{1}{K} \frac{[NH_3]^2}{[H_2]^3} \qquad (5.3.4)$$

or:

$$[N_2]_v = [N_2]_e \frac{[NH_3]^2}{[NH_3]_e^2} \frac{[H_2]_e^3}{[H_2]^3} \qquad (5.3.5)$$

where $[NH_3]/[NH_3]_e$ is by definition the efficiency η of the synthesis reaction. It is convenient to use η, as the rate depends on the concentration of ammonia (Dumesic et al., 1975b). Values of equilibrium conversion can be found in the literature (Schulz and Schaefer, 1966).

At 673 K, consider a reacting stoichiometric mixture that has reached $\eta = 0.1$, the total pressure equal to 1 atm. The degree of conversion is so small that, to an excellent approximation:

$$[N_2] \simeq [N_2]_e \qquad [H_2] \simeq [H_2]_e$$

hence:

$$[N_2]_v = \eta^2[N_2] \qquad (5.3.6)$$

From this relation, it is easy to calculate the virtual pressure of N_2 during ammonia synthesis (Table 5.3).

Similar considerations can be applied to the equilibria:

$$H_2O + * = H_2 + O*$$

and:

$$H_2S + * = H_2 + S*$$

TABLE 5.4   Turnover rates v_t at 678 K, $\eta = 0.15$, stoichiometric mixture, 1 atm

Catalyst	Particle size d (nm)	$v_t \times 10^3 \ s^{-1}$
1% Fe/MgO	1.5	1.0
5% Fe/MgO	4.0	9.0
40% Fe/MgO	30.0	35.0

which permit the variation of virtual pressures of oxygen or sulfur, over many orders of magnitude, by changing $H_2O - H_2$ or $H_2S - H_2$ mixture ratios in equilibrium with a metallic surface (Bénard, 1970).

5.32 Effect of Particle Size on Turnover Rate for Ammonia Synthesis

Small particles of metallic iron on a magnesia support were prepared by Boudart et al. (1975a). The iron particle size could be changed between 1.5 and 30 nm and determined, in part, by electron microscopy, X-ray diffraction, magnetic susceptibility, and Mössbauer spectroscopy. Agreement was satisfactory with particle size values obtained by selective chemisorption of carbon monoxide (2 Fe for 1 CO).

Two results are noteworthy. First, the turnover rate for ammonia synthesis increases by a factor of 35 as the iron particle size increases (Table 5.4). Second, a pretreatment of the iron catalyst with ammonia increases the turnover rate by only 10 percent for the larger particles, but quite appreciably (Table 5.5) for iron clusters.

TABLE 5.5   Turnover rate at 678 K for a 5% Fe/MgO catalyst, depending on the pretreatment ($\eta = 0.15$, stoichiometric mixture, atmospheric pressure) (Boudart et al., 1975b)

Pretreatment, in order	$v_t \times 10^3 \ s^{-1}$
1 Reduction by H_2-Deoxo	9.0
2 NH_3, 1 h	13.0
3 H_2-Deoxo, 20 h	9.0
4 H_2pure (diffusion through Pd) 20 h	7.0

One possible interpretation of the change in rate with particle size is a metal-support interaction that becomes more important as the size of the clusters becomes smaller. This interaction could modify the electronic structure of the metal, as will be explained in Chapter 6, accounting for the change in catalytic activity. This explanation can be rejected because the Mössbauer effect spectral parameters of metallic iron are identical within experimental error, for all of the supported samples and for bulk iron (Boudart et al., 1975a). Hence, it was concluded that *ammonia synthesis is a structure-sensitive reaction* (Boudart et al., 1975b).

Besides, it was possible, for the first time, to propose what type of surface structure was responsible for the enhanced rate of reaction. Clearly, the surface atoms or sites involved become more abundant as the cluster size grows. Since we deal with clusters, with ill-defined crystallographic planes, it is better to describe the surface structure in terms of the number of surface atoms C_i with coordination number equal to i, rather than in terms of Miller indices used for large crystals (§2.1). Dumesic et al. (1977) showed the parallel between enhanced values of turnover rates of ammonia synthesis and the following three phenomena.

The first phenomenon is ammonia pretreatment which, because of the very high virtual pressure of nitrogen, reconstructs the surface with formation of iron nitride (Dumesic et al., 1975b). It must be noted that the bulk nitride disappears when the particles are contacted with synthesis gas and that the surface reconstruction is reversibly erased by prolonged reduction in Deoxo-H_2 or palladium diffused H_2, the latter being rigorously devoid of nitrogen.

The second phenomenon is a decrease of CO chemisorption which, following Brunauer and Emmett (1940), can be attributed to less accessible surface atoms below the surface. Such atoms do not chemisorb CO. On iron surfaces, C_7 atoms are of this type (Fig. 2.2).

Finally, ammonia pretreatment brings about a decrease in the surface magnetic anisotropy of the clusters, an effect predicted by Neel in 1954. This phenomenon is observed by means of a decrease in the area of the hyperfine split lines in the Mössbauer spectrum of Fe with a simultaneous increase in the area of the peaks ascribed to superparamagnetic iron (Dumesic et al., 1975b). Thus catalysts with smaller particles exhibit an increase in superparamagnetic iron by only 3 percent following ammonia treatment, while this increase reaches 20 percent for larger particles. These observations can be explained by an increase in the surface concentration of C_7 atoms which exhibit a lower barrier to magnetic anisotropy. Since, moreover, the surface concentration of C_7 atoms increases with cluster size (Van Hardeveld and Hartog, 1969), all phenomena linked with ammonia pretreatment, as well as the effect of particle size on turnover rate, lead to the conclusion that sites associated with C_7 atoms have a higher activity in ammonia synthesis.

This conclusion is in harmony with early results of Brill et al. (1967), who claimed that (111) planes grow on an iron crystal when it is exposed to nitrogen at 673 K. The (111) plane is the plane with a high concentration of C_7 atoms as shown on Fig. 5.6. In another related study, Brill and Kurzidim (1969) observed an increase in the rate of ammonia synthesis when an iron catalyst is reduced in a stoichiometric mixture of N_2 and H_2, with the intent to expose more (111) planes than can be achieved by reduction in H_2 alone. Another related study is that of McAllister and Hansen (1973), who showed that the areal rate of decomposition of NH_3 is ten times larger on (111) planes of tungsten than on (100) or (110) planes. It will be recalled that tungsten has a body-centered cubic lattice like iron. Also, the rate of chemisorption of N_2 at 673 K is 200, or sixty times higher on Fe(111) than on Fe(100) or Fe(110) respectively (Ertl, 1981). Finally, at 20 atm the relative areal rates of ammonia synthesis on the (111), (100), and (110) planes of iron at 798 K are 418, 25, and unity respectively (Spencer et al., 1982). The conclusions of Dumesic et al. are thus supported by a considerable amount of work.

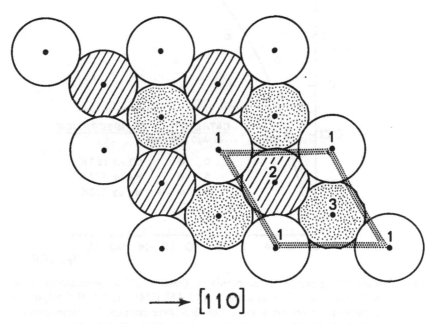

Fig. 5.6 The (111) plane of a body-centered cubic lattice

Another confirmation is found in work by Topsøe et al. (1981), who reaffirm the increase in turnover rate v_t for ammonia synthesis on iron particles of increasing size. But this was found to be true when the sites are counted by selective chemisorption of either H_2 or CO (Fig. 5.7). By contrast, when v_t is calculated by counting sites by means of high temperature chemisorption of N_2, v_t remains the same for Fe/MgO with iron particles of various sizes, as well as for unsupported promoted iron catalysts with smaller or larger particles than those of the supported samples.

Thus it appears that high temperature chemisorption of N_2 counts correctly, not the iron surface atoms, but the sites required for chemisorption of N_2, the rate of which is the rds in ammonia synthesis. It would be

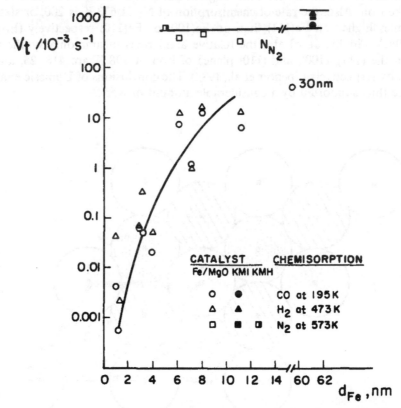

Fig. 5.7 Rate of turnover, v_t, of ammonia synthesis at 673 K, atmospheric pressure, stoichiometric mixture, $\eta = 0.15$. The KMI and KMH catalysts are promoted, unsupported iron. The sites were counted by chemisorption of H_2, CO, or N_2 as indicated.

wrong to conclude that the reaction is structure-insensitive because v_t is independent of particle size when the sites are counted with N_2. On the contrary, all the combined results of Fig. 5.7 confirm the view that a site for ammonia synthesis comprises a certain number of atoms, including C_7 atoms, forming a multiplet, or ensemble, or cluster. Perhaps these consist of the six atoms shown on Fig. 5.6 (Huang, 1981).

In conclusion, it seems that the structure-sensitivity and the surface reconstruction induced by nitrogen are due to the availability or to the creation of more active sites that include C_7 atoms.

5.4 Other Structure-Sensitive or -Insensitive Reactions

This section contains instructive details on four other reactions that are either structure-sensitive, or structure-insensitive, or both, depending on conditions.

Whenever a reaction is studied on supported metal particles of varying size, it is prudent, before one concludes that the reaction is structure-sensitive, to eliminate all "disguises" due to heat and mass transfer, or to poisoning, or to metal-support interactions.

5.41 Reaction Between H_2 and O_2 in Excess H_2 or O_2: A Possible Example, in the Latter Case, of Corrosive Chemisorption

The kinetics of the reaction:

$$H_2 + \tfrac{1}{2}O_2 \Longrightarrow H_2O \qquad (5.4.1)$$

have been studied by Hanson and Boudart (1978). This is a good test reaction when studied in excess dioxygen or dihydrogen, which must be done anyway for safety reasons when the reaction is investigated near atmospheric pressure.

The catalyst samples consist of platinum particles in the critical size range, supported on silicagel. The dispersion of the metal was determined by H_2 titration of preadsorbed oxygen (Benson and Boudart, 1965) or by selective chemisorption of H_2 (1 H per surface Pt atom). Table 5.6 contains the characteristics of the samples and the kinetic results. The table presents a clear verification of the criterion of Koros and Nowak (1967). Indeed, the second and third rows of Table 5.6 show samples with substantial changes in metal loading without a change of size or dispersion. Yet the rate constants remain the same. This indicates absence of internal or external heat or mass transfer phenomena. Poisoning or channeling in the catalyst bed can also be ruled out.

TABLE 5.6   Reaction between H_2 and O_2 on Pt/SiO_2, at 273 K

| Pt %
by weight
1 | d, particle
size nm*
2 | D
Dispersion
3 | Rate constants
10^3 (cm s^{-1}) | |
			k_1 excess H_2 4	k_2 excess O_2 5
1 3.7	6.4	0.14	3.1	2.5
2 2.3	1.4	0.62	10.7	2.8
3 0.53	1.4	0.62	10.5	3.0
4 0.38	0.9	1.0	20.0	2.0

* d was calculated by means of d (nm) $= 0.9/D$ (§1.54)

It follows that the kinetic measurements were correctly performed. Moreover, it was verified that the catalysts, after operation at 373 K, yielded identical rate data back at the starting temperature of 273 K. The measurements are thus quite reproducible and are not affected by particle growth which might be brought about by the exothermicity of the reaction.

In excess dihydrogen ($H_2:O_2 = 10:1$), the rate of reaction at the stationary state is given by:

$$v_H = k_1 S_{Pt}[O_2] \qquad (5.4.2)$$

where v_H is the rate (mol s^{-1}), k_1 the rate constant (cm s^{-1}), S_{Pt} the metal surface area (cm^2), and $[O_2]$ the concentration of O_2 (mol cm^{-3}). The rate is zero order with respect to H_2 and first order with respect to O_2. The data of column 4 show that the reaction is structure-sensitive, the rate decreasing in monotonic fashion as particle size grows.

Although macroscopic heat transfer phenomena can be ruled out by the Koros-Nowak criterion, a question can be raised concerning the evacuation of the heat of reaction from the smaller clusters. Is it possible that the temperature of the latter is higher than that of the larger ones, accounting for higher rate constants? But this complication can be disregarded, since in excess dioxygen (see below) no effect of cluster size can be detected, on the same catalysts, at similar reaction rates, and in the same reactor.

In excess dioxygen ($O_2:H_2 = 6:1$), the reaction rate is described by the equation:

$$v_{O_2} = k_2 S_{Pt}[H_2] \qquad (5.4.3)$$

where symbols have the same meaning and units as for equation (5.4.2). The only difference is that now the order is unity with respect to H_2 and zero with respect to O_2.

The first-order rate constants k_2 in Table 5.6 now show no trend and no significant range for all values of particle size. The reaction now appears to be structure-insensitive. How can this happen?

A possible explanation is related to observations such as those of Bénard (1970) showing a surface reconstruction in the presence of a strong adsorbate, e.g., the formation of an identical two-dimensional copper sulfide by adsorption of sulfur on the three low-index planes of copper. In the case of Pt/SiO_2, a *corrosive chemisorption* of oxygen could similarly erase pre-existing surface anisotropies by corrosive chemisorption of oxygen. This might then explain the absence of particle-size effect on the water synthesis reaction in excess dioxygen. At any rate, it appears that, on the same catalysts, the same reaction may exhibit structure-sensitivity or not depending on reactant ratio.

The idea of corrosive chemisorption can be compared to that of *extractive chemisorption* proposed by Kung et al. (1976) to explain the remarkable structure insensitivity in the hydrogenation on platinum of di-tert-butyl acetylene. With this substrate, steric hindrance of the substituents is such that the triple bond can make contact with surface platinum atoms only if they possess a low coordination number (edges, corners). Thus a marked effect of particle size was expected but not found. This was explained by the *extraction* of surface platinum atoms moving toward the triple bond.

5.42 Ethane Hydrogenolysis: Structure-Sensitivity, Formalism of Temkin, Multiple-Site and Alloy Effect

Kinetic results pertaining to this much-studied reaction have already been discussed (§3.22b) as a classic example of a two-step reaction on a uniform surface.

The rate equation, with fractional reaction orders:

$$v = k[C_2H_6]^\alpha[H_2]^\beta \tag{5.4.4}$$

was obtained from a two-step sequence:

$$C_2H_6 \; + \; * \xrightarrow{\;K_1\;} C_2H_x* \; | \; \frac{6-x}{2} H_2 \tag{5.4.5}$$

$$C_2H_x* + H_2 \xrightarrow{\;k_2\;} \cdots \tag{5.4.6}$$

where C_2H_x* was the most abundant reaction intermediate. The resulting equation

$$v = k_2[L][H_2] \frac{K_1[C_2H_6][H_2]^{(x-6)/2}}{1 + K_1[C_2H_6][H_2]^{(x-6)/2}} \qquad (5.4.7)$$

can be approximated by equation (5.4.4).

Besides the discomfort created by this approximation, two objections can be raised against the postulated mechanism. First, in the case of x equal to zero, it seems that a completely dehydrogenated surface species is rehydrogenated reversibly to the starting reactant. Second, it seems odd, as first remarked by Kemball (1966), that the inhibition by H_2 observed frequently for the reaction is not attributed to adsorbed hydrogen, which in fact does not appear in the mechanism.

These two objections can be obviated by assuming an irreversible adsorption of ethane on multiple sites which are covered by hydrogen as dictated by an adsorption equilbrium with H_2 (Boudart, 1972). Besides, a Temkin formalism yields directly, without approximation, the observed power rate law (5.4.4). The sequence is now:

$$2S + H_2 \overset{K_1}{=\!\!=\!\!=} 2SH \qquad (5.4.8)$$

$$* + C_2H_6 + yS \xrightarrow{k_1} C_2H_x* + ySH + \frac{6-x-y}{2}H_2 \quad (5.4.9)$$

$$C_2H_x* + H_2 \xrightarrow{k_2} \cdots \qquad (5.4.10)$$

In the first step, H_2 is chemisorbed reversibly, dissociatively, and strongly on sites S which are different from those required for chemisorption of ethane, designated by the usual symbol *. Ethane is chemisorbed by losing y atoms of H which are bound on S sites while y atoms of H are desorbed as dihydrogen. It does not follow that the site molecularity for binding C_2H_x* is high. What is assumed is that y sites of type S must be free from hydrogen for chemisorption of ethane to take place. The physical meaning of these additional S sites will be clarified later. In other words, the irreversible dehydrogenating chemisorption of C_2H_6 requires, besides the sites *, some additional sites S adjacent to the latter.

As before, it will be assumed that C_2H_x* remains the most abundant reaction intermediate, so that steps following reaction of C_2H_x* are not kinetically significant.

Let us now treat the steps (5.4.9) and (5.4.10) by the formalism of Temkin for a non-uniform surface. The only modification to the treatment of Section 4.22 is that yS is considered as a reactant, the empty sites where

C_2H_6 is adsorbed being * sites as before. Applying equation (4.2.30) yields

$$v_t = \frac{v}{[L]} = \tau \frac{k_1^0 k_2^0 [C_2H_6][S]^y[H_2]}{(k_1^0[C_2H_6][S]^y)^m(k_2^0[H_2])^{1-m}} \qquad (5.4.11)$$

It is now assumed that the fraction θ of occupied S sites is near saturation ($\theta \rightarrow 1$) so that the fraction of free sites is approximately obtained from the Langmuir adsorption isotherm:

$$K = \frac{\theta^2}{(1-\theta)^2[H_2]} \qquad (5.4.12)$$

simply as:

$$[S] = 1 - \theta \simeq K^{-\frac{1}{2}}[H_2]^{-\frac{1}{2}} \qquad (5.4.13)$$

Substitution of $[S]$ into equation (5.4.11) leads to:

$$v_t = \tau k_1^{0(1-m)}k_2^{0m}K^{(y/2)(m-1)}[C_2H_6]^{1-m}[H_2]^{m-(y/2)(1-m)} \qquad (5.4.14)$$

or:

$$v_t = k[C_2H_6]^{1-m}[H_2]^{m-(y/2)(1-m)} \qquad (5.4.15)$$

with:

$$k = \tau k_1^{0(1-m)}k_2^{0m}K^{(y/2)(m-1)} \qquad (5.4.16)$$

Comparison with the experimental rate law is straightforward:

$$\alpha = 1 - m$$
$$\beta = m - (y/2)(1 - m)$$

Thus, according to Table 3.1, for ethane hydrogenolysis on nickel:

$$\alpha = 0.7$$
$$m = 0.3$$
$$\beta = -1.2$$

hence

$$y = 4$$

The advantages of the revised treatment are that *irreversible* extensive dehydrogenation of the hydrocarbon may take place without difficulty, hydrogen adsorption is taken into account, and the fractional exponents follow without approximation (§3.22 Example 6) from the formalism of Temkin.

Let us recall again the striking results of Morikawa et al. (1937) for the hydrogenolysis of propane on nickel (§3.22b) with reaction orders $\alpha = 0.9$ and $\beta = -2.6$.

Identification with equation (5.4.8) leads to:

$$\alpha = 0.9 = 1 - m$$

hence

$$m = 0.1$$

and

$$\beta = -2.6 = m - (y/2)(1 - m)$$

hence

$$y = 6$$

The physical meaning of y is that propane is adsorbed only on sites surrounded by six available neighbors, while only four of those are required for ethane. This makes sense and was already anticipated by Kemball (1966) who, comparing the very negative (-2.6) order with respect to hydrogen in the case of propane to the less negative (-1.1) value for ethane, proposed qualitatively that this was in keeping with the need for hydrogen to liberate more sites in the case of the larger molecule.

If we replace in the above sequence the last step (5.4.10) by:

$$CH_2* + H_2 \xrightarrow{k_2} CH_4 + *$$

and now assume that the most abundant reaction intermediate is CH_2*, it is easy to show that the rate equation is not changed at all.

This example illustrates the similarity of rate equations obtained for reversible or irreversible adsorption, and for uniform or non-uniform surfaces. Rather than harp on the ambiguity of kinetics, it may be better to emphasize the value of the mechanistic ideas behind the kinetic analysis.

One of these ideas is that of a metallic multiple site (y atoms) which has been developed by Frennet et al. (1978, 1979). According to these authors, a uniform surface contains *potential sites* defined by an ensemble of metal atoms. In the surface domain where a hydrocarbon (HC) is adsorbed, three types of sites are considered: the site where HC is bound to the surface, the sites covered by HC, and the sites at the periphery of the domain which are inaccessible to another molecule of HC.

Thus, to be chemisorbed, the molecule HC requires a landing pad made up of Z adjacent and free potential sites. In hydrogenolysis of HC, both HC and H_2 compete for the same sites. When all the ensembles of Z sites are occupied, only the sites left over that are inaccessible to HC remain available to small molecules. The size of a potential site is assumed to be similar to that of a hydrogen atom (Delaunois et al., 1967).

Three types of fraction θ are then defined: θ_H relative to sites covered by H atoms, θ_{HC} relative to those covered by HC, and θ_s, the fraction of free sites. The rate of adsorption of HC is of the usual form:

$$v = k[\text{HC}]f(\theta_s) \tag{5.4.17}$$

where θ_s will be raised to the power Z to take into account the Z sites included in the landing pad. For instance, in the dissociative chemisorption of C_2H_6, one writes:

$$C_2H_6 + ZS \overset{k_1}{\rightleftharpoons} C_2H_5* + H* \tag{5.4.18}$$

and:

$$v_{\text{ads},1} = k_1[C_2H_6]\theta_s^Z \tag{5.4.19}$$

The symbol * simply denotes a chemisorbed species. Another possibility, according to Frennet et al., is *reactive chemisorption*:

$$C_2H_6 + H* + ZS \overset{k_1'}{\rightleftharpoons} C_2H_5* + H_2 \tag{5.4.20}$$

with:

$$v_{\text{ads},2} = k_1'[C_2H_6]\theta_H\theta_s^Z \tag{5.4.21}$$

But the θ's are related:

$$\theta_s = 1 - \theta_{C_2H_6} - \theta_H \tag{5.4.22}$$

If $\theta_{C_2H_6} \ll \theta_H$:

$$v_{\text{ads},1} = k_1[C_2H_6](1 - \theta_H)^Z \tag{5.4.23}$$

$$v_{\text{ads},2} = k_1'[C_2H_6]\theta_H(1 - \theta_H)^Z \tag{5.4.24}$$

In this manner, the rates depend explicitly on the fraction of sites covered with hydrogen. The model is that of a large $(Z > 6)$ multiple site the size of which depends on the size of the molecule HC. This steric effect on adsorption is superimposed on the usual energetic factors.

Martin et al. (1979) have also considered the multiplicity of the sites engaged in hydrogenolysis reactions, but the number of atoms implicated includes not only those required for adsorption but those necessary for the complete disruption of HC.

Equation (5.4.9) and the model of Frennet et al. thus differentiate between the site * on which C_2H_6 is adsorbed and y or Z other sites S. How many surface atoms are *binding* the hydrocarbon molecule?

In a study of the hydrogenolysis of n-butane and neopentane on Ir/Au/SiO$_2$ catalysts, Foger and Anderson (1980) try to explain the variation of the pre-exponential factor with alloy composition. With neopentane, the pre-exponential factor is practically independent of gold content. It is concluded that neo-pentane is adsorbed on a single iridium atom and that all of these are active, irrespective of the amount of gold in the alloy.

On the other hand, in the case of n-butane or neo-hexane, the pre-exponential factor is divided by 100 when the atomic fraction goes from 0 to 0.86. It appears that the surface density of active sites goes down when the amount of gold increases. Since activation energies and pre-exponential factors for the hydrogenolysis of these two hydrocarbons do not depend on particle size in the case of pure iridium (Foger and Anderson, 1979), the authors suggest that the active site consists of two adjacent atoms of iridium. The same conclusion was reached by Burton and Hyman (1975) in the case of ethane hydrogenolysis on nickel where two adjacent nickel atoms were pictured as the active site (see below).

In conclusion, for hydrogenolysis of hydrocarbons on metals and alloys, more work remains to be done to determine reliably the size of the active ensembles (Garin et al., 1982). Thus the quantitative conclusions of this section must be considered as tentative, although the kinetic principles seem quite secure.

The effect of alloying on the rate of hydrogenolysis of alkanes is particularly dramatic in the case of ethane on copper nickel unsupported alloys (Sinfelt et al., 1972). It was found that addition of only 7.5% Cu to nickel (region a) depresses the activity of Ni by a factor of 1,000. Further additions of Cu, up to 74% (region b) depress the activity further by a factor of 100. In region b, the activation energy E is constant but the pre-exponential factor A decreases. The opposite is true in region a. According to Burton and Hyman (1975), who calculated the surface composition of the Cu-Ni alloys, the drop in activity in region *b* is accounted for by the decrease of the surface concentration of adjacent pairs of nickel atoms. As to what happens in region *a*, the behavior of *A* and *E* suggests rather that a ligand effect of Cu on Ni may be responsible for the change of *E* and thus of the activity.

As to the change in turnover rate from one metal to another in Groups VIII and Ib, it is truly enormous for the hydrogenolysis of ethane (Sinfelt, 1972) and of neopentane (Boudart and Ptak, 1970). The variation in rate has been correlated in both cases with the percentage *d* bond character of the metallic bond following Pauling. The large variation in rate from one metal to the next for the hydrogenolysis of ethane, the structure-sensitivity of that reaction, and the large effect of alloying on activity are all in keeping with the need for the large landing pad for the hydrocarbon (see §5.5).

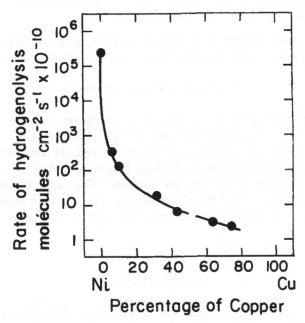

Fig. 5.8 Hydrogenolysis of ethane: effect of alloying on the areal rate of reaction (mol cm^{-2} s^{-1} × 10^{-10}), according to Sinfelt et al., 1972

TABLE 5.7 Turnover rate for hydrogenation of cyclohexene in cyclohexane at 101.3 kPa hydrogen pressure, 307 K on platinum dispersed on various supports (Madon et al., 1978)

Pt % by weight	Support	Dispersion D	v_t (s^{-1})	Particle size (nm)*
1.5	SiO$_2$	1.00	9.16	0.9
0.38	SiO$_2$	1.00	9.02	0.9
1.46	SiO$_2$	0.80	7.67	1.12
0.6	γ-Al$_2$O$_3$	0.70	8.61	1.28
2.3	SiO$_2$	0.62	8.67	1.45
0.53	SiO$_2$	0.56	8.51	1.61
0.80	SiO$_2$	0.34	8.92	2.94
1.96	η-Al$_2$O$_3$	0.23	8.21	3.91
3.7	SiO$_2$	0.14	8.62	6.43

*estimated by means of $d = 0.9/D$ (nm)

5.43 Hydrogenation of Cyclohexene in the Gas and Liquid Phases, on Large Single Crystals and Small Clusters

The reaction was studied by Madon et al. (1978) in the liquid phases near room temperature and atmospheric pressure, and in various solvents:

$$\text{cyclohexene} + H_2 = \text{cyclohexane} \qquad (5.4.25)$$

The catalysts used, the dispersion of platinum, and the values of turnover rate at specified conditions are listed in Table 5.7. The data satisfy the criterion of Koros and Nowak. Indeed, the turnover rates on the first two catalysts with a platinum loading differing by a factor of four, but identical dispersion, are identical. The same is true for the catalysts with 0.53 and 2.3% platinum. The data show no support effect and no effect of particle size. There was no detectable poisoning in this work.

The rate expression is:

$$v = k[H_2][C_6H_{10}]^0 \qquad (5.4.26)$$

with zero order with respect to cyclohexene and first order with respect to hydrogen, as shown on Figs. 5.9 and 5.10. In particular, the slope of the straight line of Fig. 5.9 gives the rate directly. The excellent reproducibility of the data, provided that the cyclohexene was free from

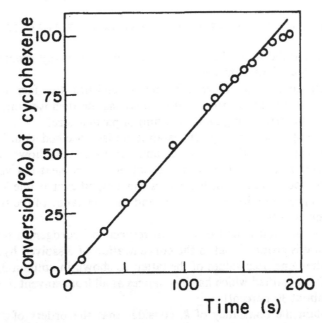

Fig. 5.9 Conversion of cyclohexene versus time on 1.5% Pt/SiO_2, $D = 1.0$, 1 atm, 275 K, in n-heptane (Madon et al., 1978).

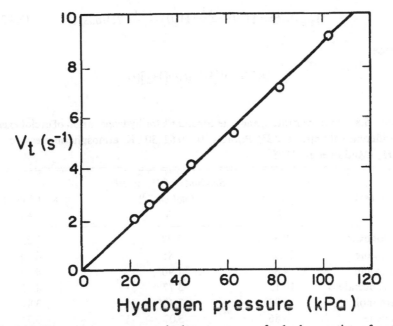

Fig. 5.10 Turnover rate v_t versus hydrogen pressure for hydrogenation of cyclohexene in cyclohexane on 2.3% Pt/SiO_2, $D = 0.62$, 307 K (Madon et al., 1978)

peroxides, and the base of the zero order behavior justify reporting the rates with three significant figures.

The last column of Table 5.7 then establishes the fact that the reaction on platinum is structure-insensitive, as the turnover rate stayed constant in the critical region (1 to 7 nm) of particle size.

The solvent effect on a given catalyst under specified conditions is shown in Table 5.8. The second column of Table 5.8 shows that the turnover rate is not constant from one solvent to the next but varies by a factor of about three, which appears quite significant if one keeps in mind the reproducibility of the results and the constancy of the turnover rate in a given solvent.

Since the reaction is first order with respect to hydrogen, it might be that the rate is proportional to the concentration of dissolved hydrogen. Indeed, when one uses values of the latter, as shown in column 3, a rate constant k is obtained which hardly changes at all from solvent to solvent (by only about 30 percent).

To explain the constancy of k, consider that the orders of reaction suggest that the measured rate is that of hydrogen chemisorption via a transition state onto a surface quasi-saturated by hydrocarbon species. The sequence of events can be represented by:

$$H_2(g) \rightleftharpoons H_2(l) \overset{K^\dagger}{\rightleftharpoons} H_2^\dagger(l) \xrightarrow{kT/h} 2H(ads, l) \qquad (5.4.27)$$

hence:

$$K^\dagger = \gamma^\dagger [H_2^\dagger](l)/\gamma[H_2](l)$$

TABLE 5.8 Turnover rate v_t and rate constant k for hydrogenation of cyclohexene in different solvents, on 2.3% Pt/SiO$_2$, $D = 0.62$, 307 K, atmospheric pressure of H$_2$ (Madon et al., 1978)

Solvent	v_t (s^{-1})	Solubility of $H_2 \times 10^6$ $(mol\ cm^{-3})$	$k \times 10^3\ cm\ s^{-1}$
1	2	3	4
Cyclohexane	8.67	3.97	3.63
n-heptane	12.57	4.85	4.30
p-dioxane	4.84	1.90*	4.23
Ethyl acetate	10.15	3.77	4.47
Methanol	7.92	3.40*	3.87
Benzene	6.86	3.07	3.71
Cyclohexene	8.40	3.57*	3.91

*calculated values

and:

$$v = \frac{kT}{h}[H_2^\dagger](l)$$

$$= \frac{kT}{h}K^\dagger\frac{\gamma}{\gamma^\dagger}[H_2](l) \tag{5.4.28}$$

The meaning of $K^\dagger$, k, T, h is as usual, while γ and $\gamma^\dagger$ are respectively the activity coefficient of dissolved hydrogen and of the transition state. The experimental rate constant becomes:

$$k_{exp} = \frac{kT}{h}K^\dagger\frac{\gamma}{\gamma^\dagger} \tag{5.4.29}$$

This result may be obtained more rigorously (Madon et al., 1978).

At a given temperature, since k_{exp} stays constant in the various media, and since the concentration of dihydrogen was used to calculate the rate constants, it must mean that the ratio of the activity coefficients $\gamma/\gamma^\dagger$ is itself independent of the nature of the solvent. In other words, in all solvents used, H_2 approaches the surface to reach the same transition state in the same environment. Thus we have:

$$v = \text{Const} \times [H_2](l)$$

as observed experimentally. There are few examples in the literature of heterogeneous catalysis where the activity has been used instead of the concentration (see Chapter 3) or where the effect of the solvent on reaction rates has been accounted for quantitatively. Yet these are important problems in theory and in practice.

The interpretation of the solvent effect on the rate of hydrogenation of cyclohexene must involve the zero order with respect to cyclohexene. This implies a surface largely covered with hydrocarbon intermediates. Equilibrium is maintained between $H_2(g)$, $H_2(l)$ and a dihydrogen precursor state immediately preceding the transition state. The precursor state is always in the same environment, no matter what the nature of the solvent may be, and its activity coefficient γ as well as that of the transition are solvent-independent.

A more radical interpretation is that of Thomson and Webb (1976) who envisage the hydrogenation of alkenes as proceeding via adsorption of dihydrogen by some unspecified hydrocarbon intermediates at the surface, followed by hydrogen transfer to the alkene itself. The difference between this mechanism and the classical mechanism of Horiuti and Polanyi (1934) is the picture of the active site, *. With both mechanisms, there must be a dihydrogen precursor state which is not affected by the solvent. Whatever the nature and the function of the hydrocarbon intermediates at the surface, they appear to be reactive, as no deactivation of the surface takes place during a run, or from run to run.

This point has been demonstrated by Hattori and Burwell (1979). These authors report that the adsorbed hydrocarbon species at the surface of platinum during the hydrogenation of cyclopropane or ethylene can be hydrogenated off the surface to form the normal products of hydrogenation, at the temperature of the hydrogenation itself, after the end of the reaction.

Many of the observations on the liquid phase hydrogenation of cyclohexene on platinum are also found in the gas phase, as reported by Segal et al. (1978), but there are some differences. In the gas phase, the reaction remains zero order with respect to cyclohexene, but the order with respect to hydrogen is fractional and varies from 0.8 at lower pressures and temperatures to 0.5 at higher temperatures and pressures. But the reaction remains structure-insensitive.

The mechanism of Horiuti and Polanyi (1934) can account for the kinetic observations. The sequence of elementary steps is the following:

$$H_2 + 2* \overset{k_1, K_1}{\rightleftharpoons} 2H* \tag{5.4.30}$$

$$R + 2* \rightleftharpoons *R* \tag{5.4.31}$$

$$*R* + H* \rightleftharpoons RH* + 2* \tag{5.4.32}$$

$$RH* + H* \overset{k_4}{\longrightarrow} RH_2 + 2* \tag{5.4.33}$$

The alkene is R. The species RH is called the half hydrogenated intermediate. If RH* is the most abundant reaction intermediate and there are only a few vacant sites between adsorbed intermediates, and if the pressure of H_2 and the temperature are relatively low, step (5.4.30) is practically irreversible and its rate is that of the reaction:

$$v = v_1 = k_1[H_2] \tag{5.4.34}$$

By contrast, at higher H_2 pressures and temperatures, step (5.4.30) becomes equilibrated. If RH* is still the most abundant reaction intermediate, the rate becomes:

$$v = v_4 = k_4[H*][RH*] = k'[H_2]^{1/2} \tag{5.4.35}$$

with:

$$k' = k_4 K^{\frac{1}{2}}[L]$$

and:

$$K^{\ddagger} = K_1^{\ddagger}[*]$$

These equations can thus account for the kinetic data. What is more important is that the latter indicate again a surface practically saturated with hydrocarbon species. This in turn may explain the observed structure-insensitivity.

Fortunately there is a direct proof of the latter in a comparison between the above data in the gas phase and those of Davis and Somorjai (1980) on a large (1 cm^2 area) single crystal of platinum exposing a stepped face (223) consisting of (111) terraces, 5 atoms wide, separated by ledges, 1 atom thick. The single crystal data were obtained at temperatures and pressures similar to those used in the previous study, so that a direct comparison can be made, as shown on Fig. 5.11. In the work of Davis and Somorjai (1980), surface structure and composition were determined by low energy electron diffraction and Auger electron spectroscopy.

At high pressure (77 torr), the agreement between the data on large crystals and small clusters is extremely good, as shown on the Arrhenius

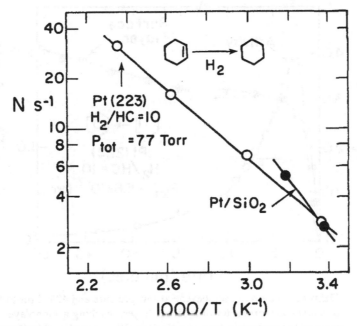

Fig. 5.11 Turnover rates v_t for the hydrogenation of cyclohexene on a large single crystal of Pt (Davis and Somorjai, 1980) and on clusters of Pt on silicagel (Segal et al., 1978)

diagram of Fig. 5.11. By contrast, at low pressure the phenomenon is totally different (Fig. 5.12). The reaction is different: it is now a *dehydrogenation* of cyclohexene to benzene, although the $H_2:HC$ ratio remains the same, equal to 10. Furthermore, after 1,000 s, less than about $2.0 \times 10^{-4} \times 10^3 = 0.2$ molecule of benzene has been produced per platinum atom. But the surface is already inactivated before it has had the time to turnover in a catalytic manner. The reaction is barely stoichiometric. It is certainly not catalytic. Deactivation is due to carbon buildup as shown by Auger electron spectroscopy. The phenomena have nothing in common with the high-pressure data. In the liquid phase, where no deactivation takes place at all, the working surface appears covered with hydrocarbon species. At low pressure, the surface coverage is too small and the adsorbed species are dehydrogenated, largely irreversibly with formation of inactive carbon layer.

The possibility of studying catalytic reactions on large single crystals with known structure and composition, at high (i.e., atmospheric) pressure, is indeed one of the most significant advances in heterogeneous catalysis. In particular, it permits a clear diagnosis of structure-sensitivity or -insensitivity to be made directly.

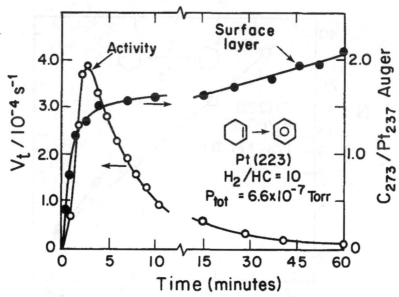

Fig. 5.12 Dehydrogenation of cyclohexene at low pressure and 423 K on Pt (223). Simultaneous formation of a carbon layer, reaching a monolayer for a ratio of the Auger electron signals C_{273}/Pt_{237} equal to 2.8 (Davis and Somorjai, 1980).

All that has been said thus far concerning the hydrogenation of cyclohexene deals with platinum. Let us now describe rapidly data on the same reaction on supported Pd catalysts (Gonzo and Boudart, 1978). The particle size was varied between 1.5 and 30 nm. Both silica and alumina were used as supports. The rate was measured both in the liquid and gas phases. In the first instance, different solvents were used. The data satisfied the Koros-Nowak criterion. The turnover rate was, at a given temperature and pressure, independent of particle size or nature of the support. As before, the order of reaction is zero with respect to cyclohexene. But now the order with respect to hydrogen is 1/2 instead of unity on platinum:

$$v = k[H_2]^{1/2} \qquad (5.4.36)$$

This may explain another difference between platinum and palladium. With the latter, the turnover rate is independent of the nature of the solvent. But now, the order of 1/2 with respect to dihydrogen suggests the following equilibrium:

$$H_2(l) + 2* \overset{K}{=\!\!=\!\!=} 2H* \qquad (5.4.37)$$

so that H_2 in solution is now in equilibrium with both gas phase and surface hydrogen so that no effect of the solvent is expected. The reaction kinetics are still described by a Horiuti-Polanyi mechanism with step (5.4.33) becoming irreversible while RH* remains the most abundant reaction intermediate.

In summary, all kinetic studies of the hydrogenation of cyclohexene on clusters of Pt or Pd, and on large single crystals of platinum, indicate that the reaction is structure-insensitive under the conditions of these investigations. These are conditions which bring about the formation of a quasi-saturated layer of active hydrocarbon species on the metal. Tentatively, the structure-insensitivity is attributed to the fact that the reaction takes place on this "metal-alkyl" surface where structural features have been essentially erased.

5.44 Gas Phase Dehydrogenation of a Secondary Alcohol: Another Structure-Insensitive Reaction with Uniform Surface Kinetics (Echevin and Teichner, 1975)

The kinetics of the reaction:

$$CH_3CH(OH)CH_2CH_3 = CH_3COCH_2CH_3 + H_2 \qquad (5.4.38)$$
$$(A) \qquad\qquad\qquad (C)$$

was studied in great detail on 18 catalysts consisting of copper (dispersion between 0.07 and 0.70) on different alumina supports. Two methods of preparation were used, with water or methanol respectively. The copper sites were counted by selective dissociative chemisorption of N_2O at 373 K (see §1.52e). The rate of turnover could be described by the classical rate expression:

$$v_t = \frac{kK_A[A]}{1 + K_A[A] + K_C[C]}$$ (5.4.39)

where k is a rate constant, K_A and K_C are adsorption equilibrium constants for alcohol and ketone respectively. This expression follows readily from the following sequence. The surface is considered to be uniform and partitioned between alcohol and ketone molecules competing for the same sites. The rate-determining step is assumed to be the dehydrogenation of the adsorbed alcohol:

$$A + * \overset{K_A}{\underset{}{\rightleftarrows}} A*$$ (1)

$$A* \overset{k}{\xrightarrow{}} C* + H_2$$ (2)

$$C* \overset{K_C}{\underset{}{\rightleftarrows}} C + *$$ (3)

The most spectacular results obtained on these eighteen catalysts is that the rate of turnover under given conditions is remarkably constant ($2.7 \times 10^{-3}\,s^{-1}$ at 451.3 K in zero order regime) on the eighteen catalysts with large changes of dispersion of copper, different methods of preparation, and different supports. Clearly, the reaction is structure-insensitive.

The study of initial rates, in the absence of reaction products, permitted obtaining a zero order kinetic regime which yielded the true surface rate constant k from:

$$v_0 = kn_s$$ (5.4.40)

where v_0 ($s^{-1}\,g^{-1}$) is the specific rate when the surface is saturated with the alcohol and n_s (g^{-1}) is the number of copper surface atoms per gram of catalyst. The activation energy was the same on all catalysts. At 451.3 K, all results for the eighteen catalysts are all on the straight line v_0 versus n_s, as befits structure-insensitive behavior. The result is:

$$v_0 = 5.01 \times 10^{-21}\, n_s \exp(-12,500/RT)$$ (5.4.41)

with the units just given, and R expressed in cal mol^{-1} K^{-1}.

On the other hand, the kinetic data also gives the adsorption equilibrium constants. They are, on all catalysts:

$$K_A = 2.81 \times 10^{-9} \exp(18,600/RT) \quad \text{torr}^{-1} \tag{5.4.42}$$

and:

$$K_C = 2.08 \times 10^{-13} \exp(30,400/RT) \quad \text{torr}^{-1} \tag{5.4.43}$$

where 18,600 and 30,400 cal mol^{-1} represent the adsorption enthalpy of the alcohol and the ketone respectively.

Another remarkable result of this work is a separate determination of K_C from adsorption isotherms obtained by means of a McBain balance. The thermodynamic value of K_C is:

$$K_C = 2.3 \times 10^{-13} \exp(30,000/RT) \quad \text{torr}^{-1} \tag{5.4.44}$$

a value which is identical to the kinetic value (5.4.43). Thus the reaction investigated appears as a model reaction from the viewpoint of structure-insensitivity and applicability of uniform surface kinetics. Clearly, the two attributes should go together. Let us note finally that Djéga-Mariadassou et al. (1982) have measured structure insensitive turnover rates for the dehydrogenation of i-propanol on zinc oxide samples with surfaces of preferred orientation. Their rate values are similar to those reported in the work of Echevin and Teichner (1975) discussed above.

5.5 CLASSIFICATION OF STRUCTURE-SENSITIVE AND -INSENSITIVE REACTIONS

We have collected in Table 5.9 only a limited number of reactions for which the classification appears clear-cut. With respect to the sensitivity to poisons, results of Barbier et al. (1979) and Leclercq and Boudart (1981) should be consulted. Otherwise, most of the other features of the table have been discussed in this chapter or elsewhere in the book.

Sometimes there exist in the literature interesting contradictions which future work should resolve. For instance, Fuentes and Figueras (1978, 1980) have found the hydrogenolysis of cyclopentane structure-sensitive on Rh/Al$_2$O$_3$ catalysts (1980), but structure-insensitive on Pd/Al$_2$O$_3$ samples (1978). Clearly, there appears to be a substantial difference in reaction mechanism on these two metals of the platinum group. It would be interesting to find out why. In fact, the apparent contradiction confirms the mechanistic interest in trying to classify reactions into two classes

TABLE 5.9 Classification of reactions: main characteristics

→ Increasing importance of effect →

	Effect of structure	Effect of alloying	Effect of poisons	Effect of nature of metal	Kinetics	Nature of bonds being activated	Multiplicity of site
$CO + \frac{1}{2}O_2 = CO_2$	CLASS 1 structure-insensitive (no effect)	minor	minor	moderate	based on uniform surface formalism	O—O	
$H_2 + D_2 = 2HD$						H—H	
$C_2H_4 + H_2 = C_2H_6$						{ H—H, C—H }	1 or 2 atoms
$C_6H_{10} + H_2 = C_6H_{12}$							
$C_6H_6 + 3H_2 = C_6H_{12}$						{ H—H, C—H }	
$C_4H_9OH \rightarrow C_3H_8CO + H_2$						O—H	
$N_2 + 3H_2 = 2NH_3$	CLASS 2 structure-sensitive (moderate effect	large	large	very large	based on non-uniform surface Temkin formalism	N—N	large multiple site
$C_2H_6 + H_2 = 2CH_4$						C—C	

↑ Increasing importance of effect →

based on structure-sensitivity and related effects: alloying, sensitivity to poisons, sensitivity to nature of the metal, and applicability of uniform surface kinetics.

There are other reactions that do not seem to be easily classified, at least not with the generalization implied in Table 5.9. Thus isotopic exchange of alkanes with deuterium seems structure-insensitive but exhibits a substantial effect of the nature of the metal on the rate (Frennet et al., 1979). Reactions between CO and H_2 may be structure-insensitive when the product is methane, but structure-sensitive when the products are higher hydrocarbons (Biloen and Sachtler, 1981). For these cases, an attempt at more sophisticated classifications is in order.

REFERENCES

Balandin, A. A. 1969. *Advan. Catal. Relat. Subj.* 19:1.

Barbier, J., Morales, A., Marecot, P., and Maurel, R. 1979. *Bull. Soc. Chim. Belg.* 88:569.

Bénard, J. 1970. *Catal. Rev.* 3:93.

Benson, J. E. and Boudart, M. 1965. *J. Catal.* 4:704.

Biloen, P. and Sachtler, W. M. H. 1981. *Advan. Catal. Relat. Subj.* 30:165.

Boudart, M. 1969. *Advan. Catal. Relat. Subj.* 20:153.

Boudart, M. 1972. *AIChE J.* 18:465.

Boudart, M., Aldag, A., Benson, J. E., Dougharty, N. A., and Harkins, C. G. 1966. *J. Catal.* 6:92.

Boudart, M., Delbouille, A., Dumesic, J. A., Khammouma, S., and Topsøe, H. 1975a. *J. Catal.* 37:486.

Boudart, M. and Ptak, L. D. 1970. *J. Catal.* 16:90.

Boudart, M., Topsøe, H., and Dumesic, J. A. 1975b. In *The Physical Basis for Heterogeneous Catalysis*, ed. E. Drauglis and R. I. Jaffee, p. 337. New York: Plenum Press.

Brill, R. and Kurzidim, J. 1969. *Colloq. Intl. Cent. Nat. Rech. Sci.* 187:99.

Brill, R., Richter, E. L., and Ruch, E. 1967. *Angew. Chem. Int. Ed. Engl.* 6:882.

Brunauer, S. and Emmett, P. H. 1940. *J. Am. Chem. Soc.* 62:1732.

Burton, J. J. and Hyman, E. 1975. *J. Catal.* 37:114.

Dalmai-Imelik, G. and Massardier, J. 1977. *Proc. 6th Intl. Cong. Catalysis*, G. C. Bond, P. B. Wells, and F. C. Tompkins, p. 90. London: The Chemical Soc.

Davis, S. M. and Somorjai, G. A. 1980. *J. Catal.* 65:78

Delaunois, Y., Frennet, A., and Lienard, G. 1967, *J. Chim. Phys.* 64:572.

Djéga-Mariadassou, G., Marques, A. R., and Davignon, L. 1982. *J. Chem. Soc. Faraday I* 78:2447.

Dumesic, J. A., Topsøe, H., Khammouma, S., and Boudart, M. 1975a. *J. Catal.* 37:503.

Dumesic, J. A., Topsøe, H., and Boudart, M. 1975b. *J. Catal.* 37:513.

Dumesic, J. A., Topsøe, H., and Boudart, M. 1977. *Proc. Nat. Acad. Sci. of USA* 74:806–810.

Echevin, B. and Teichner, S. J. 1975. *Bull. Soc. Chim. Fr.*, p. 1945.

Engel, T. and Ertl, G. 1978. *J. Chem. Phys.* 69:1267.

Engel, T. and Ertl, G. 1979. *Advan. Catal. Relat. Subj.* 28:1.

Ertl, G. 1981. *Proc. 7th Intl. Cong. Catalysis*, ed. T. Seiyama and K. Tanabe, p. 21. Tokyo: Kodansha.

Ertl, G. and Koch, J. 1973, *Proc. 5th Intl. Cong. Catalysis*, ed. J. W. Hightower, p. 969. New York: American Elsevier.

Foger, K. and Anderson, J. R. 1979. *J. Catal.* 59:325.

Foger, K. and Anderson, J. R. 1980. *J. Catal.* 61:140.

Foger, K. and Anderson, J. R. 1981. *J. Catal.* 64:448.

Frennet, A., Lienard, G., Crucq, A., and Degols, L. 1978. *J. Catal.* 53:150.

Frennet, A., Lienard, G., Degols, L., and Crucq, A. 1979. *Bull Soc. Chim. Belg.* 88:621.

Fuentes, S. and Figueras, F. 1978. *J. Chem. Soc. Faraday Trans. I* 74:174.

Fuentes, S. and Figueras, F. 1980. *J. Catal.* 61:443.

Garin, F., Aeiyach, S., Legare, P., and Maire, G. 1982. *J. Catal.* 77:323.

Gonzo, E. E. and Boudart, M. 1978. *J. Catal.* 52:462.

Goodman, D. W., Kelley, R. D., Madey, T. E., and Yates, J. T., Jr. 1980. *J. Catal.* 63:226.

Hanson, F. V. and Boudart, M. 1978. *J. Catal.* 53:56.

Hattori, T. and Burwell, R. L., Jr. 1979. *J. Phys. Chem.* 83:241.

Horiuti, J. and Polanyi, M. 1934. *Trans. Faraday Soc.* 30:1164.

Huang, K. H. 1981. *Proc. 7th Intl. Cong. Catalysis*, ed. T. Seiyama and K. Tanabe, p. 554. Tokyo: Kodansha.

Kahn, D. R., Petersen, E. E., and Somorjai, G. A. 1974. *J. Catal.* 34:294.

Kemball, C. 1966. *Discuss. Faraday Soc.* 41:190.

Koros, R. M. and Nowak, E. J. 1967. *Chem. Eng. Sci.* 22, 470.

Kung, H. H., Pellet, R. J., and Burwell, R. L., Jr. 1976. *J. Am. Chem. Soc.* 98:5603.

Ladas, S., Poppa, H., and Boudart, M. 1981. *Surf. Sci.* 102:151.

Lam, Y. L., Criado, J., and Boudart, M. 1977. *Nouveau J. Chimie* 1:461.

Leclercq, G. and Boudart, M. 1981. *J. Catal.* 71:127.

McAllister, J. and Hansen, R. S. 1973. *J. Chem. Phys.* 59:414.

Madon, R. J., O'Connell, J. P., and Boudart, M. 1978. *AIChE J.* 24:104.

Mars, P., Scholten, J. J. F., and Zwietering, P. 1960. In *The Mechanism of Heterogeneous Catalysis*, ed. J. H. De Boer, p. 66. Amsterdam: Elsevier.

Martin, G. A., Dalmon, J. A., and Mirodatos, C. 1979. *Bull. Soc. Chim. Belg.* 88:559.

Morikawa, K., Trenner, N., and Taylor, H. S. 1937. *J. Am. Chem. Soc.* 59:1103.

Morikawa, K. amd Ozaki, A. 1971. *J. Catal.* 23:97.

Neel, L. 1954. *J. Phys. Radium* 15:225.

Ponec, V. and Sachtler, W. M. H. 1973. *Proc. 5th Intl. Cong. Catalysis*, ed. J. W. Hightower, p. 645. New York: American Elsevier.

Sachtler, W. M. H. and Van der Plank, P. 1969. *Surf. Sci.* 28:62.

Scholten, J. J. E., Zwietering, P., Konvalinka, J. A., and De Boer, J. H. 1959. *Trans. Faraday Soc.* 55:2166.

Schulz, Von G., and Schaefer, H. 1966. *Ber Bunsenges Phys. Chem.* 70:21.

Segal, E., Madon, R. J., and Boudart, M. 1978. *J. Catal.* 52:45.

Sinfelt, J. H. 1972. *J. Catal.* 27:468.

Sinfelt, J. H., Carter, H. L., and Yates, D. J. C. 1972. *J. Catal.* 24:283.

Somorjai, G. A., Castner, D. G., and Blackadar, R. L. 1980. *J. Catal.* 66:257

Spencer, N. D., Schoonmaker, R. C., and Somorjai, G. A. 1982. *J. Catal.* 74:129.

Temkin, M. I. and Pyzhev, V. 1940. *Acta Physico Chim.* 12:327.

Thomson, S. J. and Webb, G. 1976. *J. C. S. Chem. Comm*, p. 256.

Topsøe, H., Topsøe, N., Bohlbro, H., and Dumesic, J. A. 1981. *Proc. 7th Intl. Cong. Catalysis*, ed. T. Seiyama and K. Tanabe, p. 247. Tokyo: Kodansha.

Van Hardeveld, R. and Hartog, F. 1969. *Surf. Sci.* 15:169.

Vannice, M. A. 1976. *J. Catal.* 44:152.

Weissman-Wenocur, D. L. and Spicer, W. E. 1982. *Surf. Sci.*, in press.

Wong, S. S., Otero-Schipper, P. H., Wachter, W. A., Inoue, Y., Kobayashi, M., Butt, J. B., Burwell, R. L., Jr., and Cohen, J. B. 1980. *J. Catal.* 64:84.

NOTE: The absence of a temperature rise in small metallic clusters on which exothermic catalytic reactions are taking place has been considered in some detail by W. L. Holstein and M. Boudart, *Latin Am. J. Chem. Eng. and App. Chem.* 13 [1983]: 107. It is due essentially to the relatively slow turnover rates (1 s^{-1}) of catalytic reactions in porous media.

The low-pressure oxidation of carbon monoxide on small particles of palladium has been reinvestigated to clarify some of the conclusions discussed on pp. 159–60. Adsorption of CO was studied on palladium particles that were vapor-deposited on various alumina supports. On the smaller particles, carbon deposition occurred by disproportionation ($2CO \rightarrow CO_2 + C_{ads}$). But on larger particles, CO remained undissociated. The turnover rate for the $CO-O_2$ reaction on palladium was shown to be independent of particle size provided that the number of surface atoms available for the reaction is estimated correctly. The number varies for the small particles, as at low temperatures the deposited carbon blocks a fraction of the available sites, while the latter become available at higher temperatures, at which surface carbon reacts away with O_2 during the $CO-O_2$ reaction. Thus, while the main reaction (CO oxidation) appears structure-insensitive, the side reaction (disproportionation of CO) is structure-sensitive. In conclusion, no special explanation (as that given on p. 160) is required. The work summarized here is by S. Ichikawa, H. Poppa, and M. Boudart, in *ACS Adv. Chem. Series*, no. 248, ed. T. Whyte, Washington, D.C., 1984, chap. 23.

Chapter 6

PARASITIC PHENOMENA

6.1 MASS AND HEAT TRANSFER

The necessity to eliminate the existence of substantial gradients of temperature and concentration in kinetic studies of heterogeneous catalytic reactions has been emphasized in the preceding chapter. To do justice to these effects requires the well-developed science of chemical reaction engineering (Denbigh and Turner, 1970; Froment and Bischoff, 1979). Our purpose here is simply to attract attention to some consequences of these effects and suggest ways to avoid them in the laboratory.

The nature of the problem is suggested by Fig. 6.1, which represents schematically a typical grain, of the order of 1mm in size, of a sup-

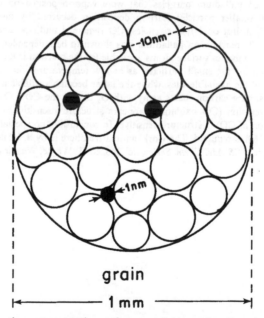

grain

|← —————— 1 mm —————— →|

Fig. 6.1 Schematic representation of a catalyst grain (1mm) with support particles (10 nm) and metallic clusters (1 nm)

ported metal catalyst. The grain consists of particles of support, amorphous or polycrystalline, say 10 nm in size. The pores of the grain contain metallic clusters of the order of 1 nm (see Fig. 1.5). It is thus necessary to distinguish between heat and mass transfer, first between the bulk of the interparticle fluid and the surface of the grains, and second, within the pores of the grains themselves. In other words, there can be *external* and *internal* gradients with respect to the grains.

6.11 External and Internal Gradients

Consider a porous catalyst grain, spherical in shape, 1 mm in size, and let us limit our attention to concentration gradients, the situation with respect to temperature gradients being analogous.

Two concentration gradients are possible (Fig. 6.2). The first is due to inadequate *external diffusion* in the interparticle fluid between the bulk of the fluid and the surface of the grain. The second is due to insufficent *internal diffusion* inside the pores of the grain.

To decrease external gradients, the relative velocity of fluid (gas or liquid) and grains must be increased. In a liquid phase slurry reactor,

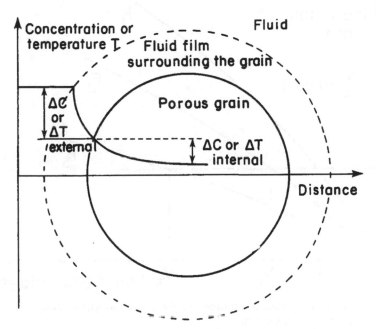

Fig. 6.2 External and internal gradients of temperature and concentration, around and inside a porous catalyst grain, for an endothermic reaction

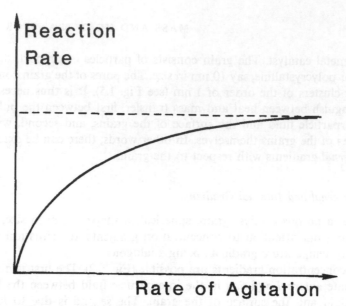

Fig. 6.3 Experimental verification of the absence of external
gradients in a slurry reactor

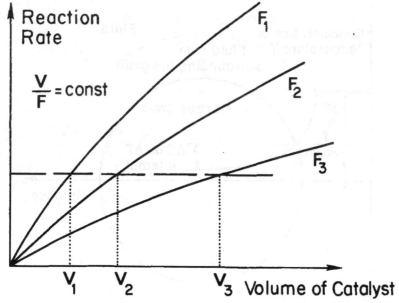

Fig. 6.4 Absence of external diffusion in a fixed bed catalytic reactor
V: volume of catalyst
F: volumetric flow rate
The rate is the same at equal values of space-time (de Mourgues et al.,
1965)

adequate agitation will not only decrease external gradients but will also keep the catalyst grains in suspension. A frequently used criterion of negligible external gradients rests on the observation that the measured rate of reaction ceases to increase any further at sufficiently high rates of agitation (Fig. 6.3). The asymptotic value of the reaction rate corresponds to external gradients that are small enough to be neglected. In the case a fixed bed, the control of absence of external diffusion is based on the same principle. Fig. 6.4 shows how reaction rate changes with volume of catalyst V at various values of the volumetric flow rate F. The reaction rate should be the same at constant value of space time V/F (see §1.27). To keep V/F constant, as F decreases, the thickness of the bed is decreased or the catalyst in the bed is diluted by a solid inert material.

The above two criteria are well-known and generally recommended. Yet there are problems in their application as evidenced in the two following examples.

6.12 Example 1: Catalytic Dehydrogenation of Propan-2-ol in the Liquid Phase, Catalyzed by Suspended Nickel (Mears and Boudart, 1966)

$$CH_3CHOHCH_3 = CH_3COCH_3 + H_2$$

The reaction rate was measured by the amount of hydrogen made per unit time. A curve similar to that of Fig. 6.3 was obtained. Does the latter mean that external mass transfer was eliminated by adequate agitation? If so, what about internal mass transfer inside the catalyst grains? How can one make sure that they are or are not important? A commonly used solution, namely to change the size of the grains, may not always be practical.

An alternative procedure is to change the amount of active surface in the porous grains, to measure the number of sites for each sample, and to obtain the turnover frequency. In the case under discussion, this was done by adding textural promoters (chromium salts) during the preparation of the catalyst so as to increase the specific surface area of nickel from 10 to 70 $m^2 g^{-1}$ as measured by the selective chemisorption of a fatty acid on nickel. Yet the turnover frequency remained essentially constant for all catalyst samples. Thus the true reaction rate was indeed measured undisguised by external or internal gradients.

It must be concluded that the increase in reaction rate with agitation (Fig. 6.3) is due, at least in part, by improved contacting between fluid and solid, the latter being kept in suspension by more intensive mixing.

At any rate, the best way to check the absence of mass and heat transfer or of other artifacts due to contacting is to use the experimental criterion of Koros and Nowak already discussed in Chapter 5. The procedure described above is an example of that criterion. Its use is particularly easy in the case of metal-supported catalysts.

6.13 Example 2: Dehydrogenation of Cyclohexane on a Pt/Al₂O₃ Reforming Catalyst (Khoobiar et al., 1965)

This kinetic study was carried out with a relatively large fixed bed under conditions close to those used in industrial catalytic reforming: total pressure ~ 2 MPa, hydrogen to hydrocarbon ratio about six, entrance temperature 700 K. The reaction is very endothermic so that it is limited by equilibrium with smaller yields of benzene as the temperature goes down. For this reason, it was checked by thermocouple readings in an axial well that no axial temperature gradient existed under reaction conditions. To improve heat transfer, the bed of catalyst grains was diluted by grains of alumina, which were verified to be inactive for the reaction studied. Then the conversion was shown to be practically independent of feed rate at values of space time that were kept the same by changing the dilution of the bed. Thus it appeared that external transport phenomena could be neglected.

Yet a calculation of equilibrium conversion at various temperatures revealed that the amount of benzene obtained exceeded the thermodynamic value corresponding to a surface temperature T_s calculated by a correlation of heat transfer in packed beds. Indeed the correlation indicated a value of T_s substantially lower than the temperature in the bulk of the fluid between catalyst grains.

Hence a paradox: the experiment failed to reveal any external temperature gradient, whereas calculation indicated sizable gradients that in fact led to values of T_s for which the permissible equilibrium value of the product was inferior to that measured experimentally.

To explain this situation, Khoobiar et al. postulated that platinum produced atomic hydrogen that was responsible for the reaction, either at the surface of the grains of alumina that did not contain platinum, or in the gas phase, through an unspecified chain mechanism.

With such a chain reaction, the main part of the heat of reaction need not be supplied to the Pt/Al₂O₃ grains but to the space in the bed surrounding these grains. This would account for the lack of experimental detection of external gradients in spite of the calculated existence of these, if the reaction did take place only in the Pt/Al₂O₃ grains.

This bold new hypothesis of Khoobiar et al. has been verified subsequently for different catalysts, reactions, and conditions (Khoobiar,

1964; Boudart et al., 1969; Lacroix et al., 1981). The phenomenon has been called hydrogen spillover.

Nevertheless, it does not appear necessary to invoke hydrogen spillover to explain the observations of Khoobiar et al. discussed above. Indeed, Chambers and Boudart (1966) calculated by means of a different correlation of heat transfer in packed beds that external temperature gradients were too small to have been detected experimentally or to lead to such low values of T_s that the measured yield would have exceeded the permissible equilibrium value. The problem is that in a laboratory reactor the size of the grains and the flow rates are smaller than in large-scale reactors, so that the Reynolds number in the bed is very small. As a result, heat and mass transfer correlations in packed beds are not reliable, and varying flow rate has a small effect on heat and mass transfer coefficients.

Laboratory reactors can be designed to operate under so-called *gradientless* conditions. They are available commercially. But even with those, it may be prudent to verify the absence of all gradients by performing a Koros-Nowak test whenever feasible.

6.14 Apparent Activation Energy in the Regime of Internal Diffusion

The problem of diffusion and reaction in porous catalytic media is a classic one in chemical reaction engineering. It has been treated experimentally and theoretically over a period of thirty years with a great deal of refinement. Its elementary presentation here may not be out of place, as it seems to remain unappreciated even today by chemistry-oriented kineticists. According to a well-known Einstein relation, the average distance L over which a particle with a diffusion coefficient D travels in time t is given by:

$$L^2 = 2Dt \qquad (6.1.1)$$

When reactants must diffuse through a reactive medium, liquid or porous, there exists a *penetration effect* which is controlled by the relative values of the diffusion coefficient and the reaction rate. This phenomenon must *always* be taken into account in any situation dealing with porous catalysts.

Consider a porous layer of thickness L and unit surface area. Assume a first-order reaction with a rate constant k and a reactant A with concentration $[A]_0$ at the external surface of the layer. If diffusion rates could maintain this concentration all through the layer, the rate in the layer would be:

$$(d\xi/dt)_{ideal} = k[A]_0 L \qquad (6.1.2)$$

But in reality, diffusion gradients will be set up in the layer, small as they may be. If these are large so that $[A] = 0$ at layer depths smaller than L, everything happens as if the rate in the layer were that given by a reactant at $[A] = [A]_0$ over an effective thickness $\bar{L}$ of the layer:

$$(d\xi/dt)_{ideal} = k[A]_0 (\bar{L}^2)^{1/2}$$

$$= k[A]_0 \left(\frac{2D}{k}\right)^{1/2} \qquad (6.1.3)$$

or:

$$(d\xi/dt)_{exp} = k_{exp}[A]_0 L$$

with:

$$k_{exp} = (2\,Dk)^{1/2}(1/L) \qquad (6.1.4)$$

Hence, if we neglect the temperature dependence of D as compared to that of k, the activation energy E_{exp} in the case of severe internal diffusion limitation is half the kinetic value E_{ideal}

$$E_{exp} = \frac{E_{ideal}}{2} \qquad (6.1.5)$$

Equation (6.1.5) explains many literature observations showing a progressive decrease of the activation energy as temperature increases. Note in particular that even in the regime of severe limitation of the rate by internal diffusion, the apparent activation energy can still have a substantial value which far exceeds that expected for a diffusion coefficient even in the liquid phase. As an example, it is not excluded that internal diffusion accounts for Arrhenius diagrams similar to that sketched in Fig. 6.5, as shown in a paper by Taghavi et al. (1978). A situation of the type of Fig. 6.5 is also another case of a compensation effect between activation energy and pre-exponential factor.

An experimental criterion to ascertain the absence of internal diffusion consists of decreasing the size of the catalyst grains and checking that the specific reaction rate does not change. The criterion may be difficult to apply (e.g., in trickle bed reactors, or if the grains are very small to start

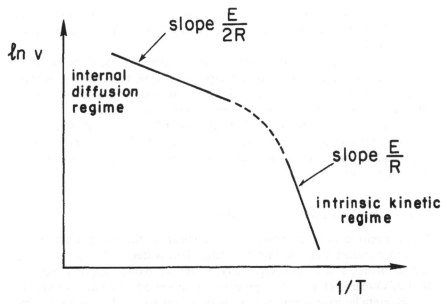

$$\text{slope } \frac{E}{2R}$$

$\ln v$

internal
diffusion
regime

$$\text{slope } \frac{E}{R}$$

intrinsic kinetic
regime

$1/T$

Fig. 6.5 Values of the activation energy in the kinetic and internal diffusion
regimes

with). It may fail if the pore distribution is bimodal with mesopores and macropores. Reducing the grain size may eliminate the influence of diffusion in the macropores without modifying a limitation due to diffusion in the mesopores.

Again, a safe universal criterion is that of Koros and Nowak. A *caveat* exists in the case of an exothermic reaction. As discussed by Madon and Boudart (1982), a coincidence could exist such that the increase in temperature inside the grain would lead to an increase in the rate constant that would be compensated exactly by the drop in the rate due to exhaustion of reactant along the pores. Such an accidental compensation should be lifted by repeating the Koros-Nowak test at another temperature.

6.2 CLASSIFICATION OF METAL-SUPPORT EFFECTS (BOUDART, 1979)

In the examples discussed in Chapter 5, the turnover rate for certain reactions on a given metal was found to remain the same on different samples of the catalyst. The absence of effects of structure or texture can be verified by changing metal cluster size in a critical range (< 5 nm), or by changing the support, or by changing the metal loading at constant dispersion, or by comparing data obtained with the supported catalysts and with macroscopic single crystals exposing clean and well-defined surfaces of different orientation.

In such situations, the absence of a support effect can be demonstrated. But if turnover rates change from sample to sample, what should be concluded? For instance, if a cluster-size effect is observed, is this due to a change of surface structure or to a metal-support interaction? If the latter prevails, what is the nature of the interaction? In what follows, it will be seen that several parasitic effects can occur which can be ascribed to apparent metal-support interactions. But these effects are interesting in their own right.

6.21 Incomplete Reduction of the Metal

A first eventuality is the absence of reduction of the catalyst precursor to the zerovalent state, in part or totally. This is frequently so with iron, cobalt, nickel, and copper, especially on certain oxide supports. For instance, Dumesic et al. (1975) prepared a collection of catalysts consisting of metallic iron supported on magnesium oxide. It was shown by Mössbauer effect spectroscopy that the fraction of iron reduced to the metallic state varied relatively little, between 0.4 and 0.7 as the amount of iron in the support increased from 1 to 40 percent by weight. Correspondingly the fraction of unreduced iron (Fe^{2+}) in solution in the magnesium oxide changed relatively little also. This phenomenon can perhaps be explained by an oxidation-reduction involving clusters of Fe^{2+} in the support and the iron itself.

Another example is that of nickel in zeolites (Briend-Faure et al. 1978) or iron in zeolites (Delgass et al., 1969) where reduction of the metal is particularly arduous. These phenomena, of course, can be considered *lato sensu* as a form of metal-support interaction which can even be called strong metal-support interaction, although that expression has been used to describe other effects (Tauster et al., 1978), as discussed below.

6.22 Support-Induced Cluster Size

Because of its texture, a support can induce and maintain a particle size of the metal. This has been exploited to grow 1 nm clusters of iron supported on alumina (Fujimoto and Boudart, 1979). In that example, precipitation of the iron hydroxide precursor takes place on the walls of pores which limit the growth of iron clusters during reduction. Again, reduction to the zerovalent state affects only about half of the iron. It appears that the remaining FeO is strongly interacting with the support.

A more subtle effect of the texture of the support has been reported by Vanhove et al. (1979, 1980) in a study of hydrocarbon synthesis from CO and H_2 on cobalt in various supports. It appears that the selectivity to hydrocarbons depends on porosity as a result of a cage effect which controls chain length. This is a variation on the theme of shape selectivity.

6.23 Epitaxial Growth of Metals on Supports

A choice example of this effect deals with the preparation of silica-supported nickel, starting with nickel antigorite (Dalmai-Imelik et al., 1974). This compound, $Ni_3OH Si_2O_5(OH)_3$, has a layered structure. Upon reduction by hydrogen, an endotactic growth takes place that produces nickel platelets with (111) or (110) orientations parallel to the support layers. Epitaxial growth is controlled by this orientation with the support. A process such as this one requires close matching in the lattice parameters of support and metal. Sometimes it can be promoted by the presence of a gas (Pashley, 1956).

6.24 Support Effect on the Morphology of Metallic Particles

A related effect can affect the *shape* (morphology) of metal particles on supports so that either small spheres tangential to the surface or rafts, i.e., layers one or two atoms thick, can develop (Sinfelt, 1979).

The effect of support on the shape of metal clusters in contact with it is little understood beyond the knowledge that it is determined by interfacial contact energies and surface energies as controlled by adsorbed gases. These phenomena are very poorly documented at the present time. In the work of Wong et al. (1980), the turnover rate of platinum in the hydrogenation of cyclopropane to propane remains constant as the percentage of platinum exposed increases to 100 percent. For this value of dispersion, the turnover rate for Pt/Al_2O_3 keeps the same value but increases by a factor of two for Pt/SiO_2. As other support effects discussed in this chapter appear unlikely to explain the difference between alumina and silica as a support for platinum in a hydrogenation reaction, it may be speculated that the shape of the clusters is different on silica (e.g., spherical) or on alumina (e.g., rafts) as a result of the fact that the platinum-support interfacial energy is larger with alumina than with silica.

6.25 Contamination of the Metal by the Support

This occurs frequently. Here are a couple of examples. In reducing Pt/Al_2O_3 samples at high temperatures in hydrogen, Den Otter and

Dautzenberg (1978) observed a decrease in the ability of the metal to chemisorb hydrogen. This was attributed to the reduction of Al_2O_3 to Al made possible by the formation of an alloy between Al and Pt. In other cases, an impurity in the support, e.g., iron or sulfur, may contaminate the metal. The difficulty of detection of the phenomenon can be due to the fact that the metal may be required for the impurity of the support to be reduced so that a blank reduction with the support alone is negative. Examples are: iron impurities on silica-gel as contaminants of palladium (Ladas et al., 1978; Fuentes and Figueras, 1978) and sulfur coming from sulfate impurities on alumina as contaminants of platinum (Maurel et al., 1975).

6.26 Bifunctional Catalysis

The most famous example is the reforming of naphtha on Pt/Al_2O_3 where two catalytic functions co-exist side by side: an acid function on alumina favoring the formation of carbonium ion intermediates intervening in skeletal rearrangements and cracking of hydrocarbons, and a metallic function favoring dehydrogenation (e.g., of alkanes and cycloalkanes). The cooperation between the two functions and their close proximity are essential to the operation of this mode of metal-support interaction.

Thus in alkane isomerization (see Chapter 5), the alkane is first dehydrogenated on platinum with formation of an alkene which migrates to an acid site of the alumina where it is rearranged. The isoalkene then desorbs and moves back to the metal for hydrogenation to the final isoalkanes. Since the intermediate must diffuse from one function to the other, a diffusional limitation might be expected when the distance between the two functions is made longer. This effect has indeed been demonstrated (Weisz, 1975).

An inverse type of bifunctionality has been invoked by Schlatter and Boudart (1972) in the hydrogenation of methylcyclopropane on Pt/SiO_2. The reactant is first isomerized to n-butene on acidic aluminum impurities on the silica surface. The intermediate then moves over to the metal where it is hydrogenated to n-butane. This mode of bifunctionality is selective, as direct hydrogenation of methylcyclopropane on the metal yields isobutane instead of n-butane.

In summary, bifunctional catalysis involves a subtle mode of interaction between metal and support which may lead to different products, depending on whether the reaction starts on the metal or on the support.

6.27 Spillover Phenomena

The most striking demonstration (Khoobiar, 1964) of the phenomenon is the reduction of yellow WO_3 to blue H_xWO_3 ($x \sim 0.35$) by H_2 at room temperature when a WO_3 powder is mixed mechanically with a Pt/Al_2O_3 powder. By contrast, a reduction by H_2 of WO_3 alone to lower oxides requires temperatures higher than 500 K.

In fact, for the Khoobiar reaction to take place at room temperature, water is an essential co-catalyst (Benson et al., 1966), whether it is introduced with the hydrogen or made by the latter at the surface of oxidized platinum.

The kinetic role of water was confirmed by Boudart et al. (1969) in a study of isotope effects where H_2O or D_2O as well as H_2 or D_2 were used in the platinum catalyzed reduction of WO_3. It was also shown that the rate did not depend on the amount of platinum, suggesting that the rate-determining step does not take place at the surface of the metal. The rate is not limited either by the mobility of hydrogen atoms or protons in the H_xWO_3 phase, as shown by an NMR study of Vannice et al. (1970).

The mechanism of hydrogen spillover thus starts at the metal surface where dihydrogen dissociates. Hydrogen atoms then spill over to the non-metal phase. How does this migration take place?

Levy and Boudart (1974) have studied that question by replacing water as a co-catalyst by other proton acceptors (alcohols). The rate of spillover could be correlated to the proton affinity of the co-catalysts. It appears that the co-catalyst is protonated at the surface of the metal with transfer of the valence electron of hydrogen to the conduction band of the metal. An ion-electron pair moves over to the surface of WO_3 where the protonated co-catalyst gives up its proton as hydrogen in a rate-determining step which is the faster the lower the proton affinity of the co-catalyst.

In catalysis, hydrogen spillover has been invoked for the hydrogenation of ethylene by Carter et al. (1965) and by Teichner et al. (1977). The latter authors have demonstrated the phenomenon by means of an ingenious device in which a support is first contacted with hydrogen in the vicinity of a supported platinum sample. The latter is then removed

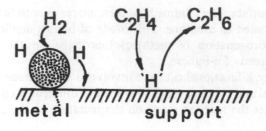

and the support activated by spillover is used in the reaction, as shown in the above diagram.

An inverse spillover effect was first postulated by Taylor (1960), under the name of *porthole effect*. The latter was invoked by Fujimoto and Toyoshi (1981) in a study of the dehydrogenation of pentane on carbon with metals supported on it. The phenomenon is represented schematically. It appears that hydrogen formed in the dehydrogenation of pentane

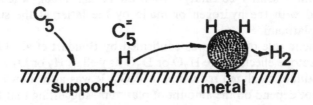

on carbon blocks the sites unless it is evacuated through the porthole, i.e., a metal particle in the support. This evacuation is an inverse spillover where hydrogen migrates from the support to the metal. The latter plays no direct role in the reaction of dehydrogenation.

Note that metallic impurities are known to catalyze the gasification of carbon by a number of reactants (O_2, H_2, H_2O, CO_2). Thus steam gasification of coal can be catalyzed by potassium carbonate. The mechanism of these catalytic processes could involve spillover or adlineation (see below). An inverse spillover mechanism in which carbon atoms are migrating from the carbon to the metal is supported by kinetic data for the gasification of carbon catalyzed by platinum (Holstein and Boudart, 1981). Similarly, in the oxidation of graphite catalyzed by iridium and rhodium, Baker and Sherwood (1980) have shown by direct observation in a controlled atmosphere transmission electron microscope how the reaction takes place as long as an interfacial contact is maintained between the carbon and the metal. The importance of this interface was stressed by Schwab many years ago: the effect was called adlineation (see Boudart et al., 1969).

6.28 Schwab Effect of the First Kind

In a number of papers, Schwab and his co-workers (1946, 1950, 1957, and 1959) have formulated a concept according to which catalytic activity of a semiconductor is affected by its contact with a metal. This will be called the Schwab effect of the first kind. An effect of the second kind is the inverse of the first, in which the catalytic activity of a metal is modified by its contact with a non-metallic support.

An example of the first kind of effect is zinc metal with a surface layer of zinc oxide. As a result of the contact between the metal and the oxide, a modification of the density of charge carriers is set up in the so-called Schottky layer of the semiconductor. If the rate of a reaction catalyzed by the semiconductor depends on the charge density at the surface of the latter, a Schwab effect of the first kind would be expected.

6.29 Schwab Effect of the Second Kind

This is the only metal-support interaction which may be called as such in a strict manner. All other effects (incomplete reduction, texture, epitaxy, impurities, bifunctionality, spillover) can be considered as parasitic or indirect. In this second effect of Schwab, charge transfer or polarization at the interface between a very small metal cluster (1 nm), and a support

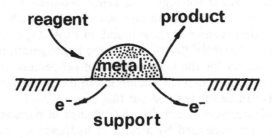

can modify the electron density of cluster atoms with a consequent modification in their catalytic activity. A strong metal-support interaction of this sort has been studied extensively in the Pt/Y-zeolite system.

This catalyst was first described by Rabo et al. (1965). It was prepared by exchanging sodium ions from a Y-zeolite with $Pt(NH_3)_4^{2+}$. It was reported that after reduction, the metal was atomically dispersed, i.e., present as individual zerovalent atoms, and that the catalyst was more resistant to sulfur poisoning in hydrocarbon reactions than other platinum-supported catalysts. Later studies of a similar catalyst (Linde SK-200) could not confirm the atomic dispersion by X-ray absorption edge measurements (Lewis, 1968) or selective chemisorption of H_2 and O_2 (Boudart, 1969). Rather, it was suggested that the dispersion of the metal was only 0.3 to 0.5, indicating a particle size too big to fit in the supercages of the zeolitic structure (~ 1.3 nm in diameter).

A new preparation of Pt/Y-zeolite was reported by Dalla Betta and Boudart (1973). The size of the platinum clusters was determined by an indirect method which suggested about six atoms per cluster. But to obtain these small clusters, it was necessary to first destroy the $Pt(NH_3)_4^{2+}$ cation by oxidation and then to reduce the metal in the presence of the least possible amount of water vapor, as the latter was shown to bring about cluster growth. In fact, probably because of that restriction, the preparation succeeded only for small quantities of catalyst (0.5 g). It may well be that this explains the failure of the commercial catalyst (SK-200) to exhibit small clusters of metal. The size of the clusters has now been confirmed by de Ménorval et al. (1982), who conclude that the clusters contain between four and eight atoms of platinum. This conclusion was reached by a study of the NMR chemical shift of xenon adsorbed in the Pt-Y zeolite as increasing amount of hydrogen were adsorbed on the clusters.

Tables 6.1 and 6.2 show turnover rates of platinum in these clusters for the isomerization (N_I) and hydrogenolysis (N_H) of neopentane, and for the hydrogenation of ethylene (N_E). The salient result is that values of turnover rates are much higher on acidic zeolites Pt/CaY, Pt/MgY, Pt/RE (RE $\equiv$ rare earth) than on non-acidic zeolites Pt/NaY for which turnover rates are the same as those found on Pt/Al_2O_3 or Pt/SiO_2. The enhancement is substantial (factor of 5) in the hydrogenation of ethylene, but it is even higher for the isomerization of neopentane (factor of 60). Yet the latter cannot form a carbonium ion of the acidic support so that bifunctionality can be safely ruled out this case.

The rate enhancement is not due to a change in particle size, as the size of the clusters measured by a ratio of hydrogen to total platinum (Table 6.2) is the same for non-acidic and acidic zeolites. Rather, it appears

TABLE 6.1   Turnover rate for isomerization N_I and hydrogenolysis N_H of neopentane at 545 K, atmospheric pressure, and $H_2:HC = 10$.

Catalyst	Dispersion Pt	$N_I \times 10^4 \, s^{-1}$	$N_I \times 10^4 \, s^{-1}$
4.9% Pt/CaY	1.0	54.00	120.0
1.96% Pt/η-Al$_2$O$_3$	0.64	0.86	2.7
1.96% Pt/η-Al$_2$O$_3$	0.076	1.70	4.2

TABLE 6.2 Turnover rate for hydrogenation of ethylene (N_E) at 189 K, atmospheric pressure, 23 torr C_2H_4, 152 torr H_2 and 585 torr He.

Catalyst	H/Pt (total)	$N_E \times 10^3 s^{-1}$
0.54% Pt/NaY	1.41	5.34
0.59% Pt/CaY	1.34	25.00
0.60% Pt/MgY	1.40	23.30
0.50% Pt/REY*	1.40	20.30
0.53% Pt/SiO$_2$	0.56	6.31

*RE: rare earth

that the strong electrostatic field gradients in the acidic zeolite cages (Rabo et al., 1965) might polarize the clusters, so that the clusters in the acid zeolites appear to be *electron deficient*. Indeed, these platinum clusters behave catalytically more like iridium, which is the element to the left of platinum in the periodic table, since iridium is more active than platinum in the two reactions of neopentane.

An electron deficiency suggested by chemical analogy is also in line with the smaller amount of oxygen chemisorbed by these clusters as well as their stronger resistance to sulfur poisoning (Rabo et al., 1965).

Spectroscopic studies have confirmed the postulated electron deficiency of these clusters. Infrared spectroscopy of adsorbed CO (Gallezot et al., 1977), X-ray photo-electron spectroscopy (Ioffe et al., 1977; Vedrine et al., 1978; Foger and Anderson, 1978), electron spin resonance of adsorbed molecules forming radical cations or anions, and finally X-ray absorption spectroscopy (see below) all confirm the earlier chemical intuition.

Gallezot et al., (1979) have studied on several Pt/Y samples the shift in the L_{III} absorption edge of platinum as well as the change in area of the threshold absorption peak ("white line") due to the transition from $2p$ core levels to $6s$ and $5d$ unoccupied levels in the valence band of the metal. The edge and the white line are represented schematically on Fig. 6.6. A shift in the absorption edge characterized by the position of its inflection point i toward higher energies indicates higher positive charges on the metal. Similarly, an increase in the area of the white line corresponds to a decrease in the population of $6s$ and $5d$ levels in the valence band. Both effects thus indicate a decrease in the electron density of the metal. The results show meaningful differences between Pt/NaY and Pt/acidic Y samples: the absorption edge is displaced toward higher energies in acidic zeolites. This is also the case when the clusters are covered with oxygen, so that the metallic structure is destroyed (Gallezot et al., 1978) and a platinum oxide is formed with electron deficiency of

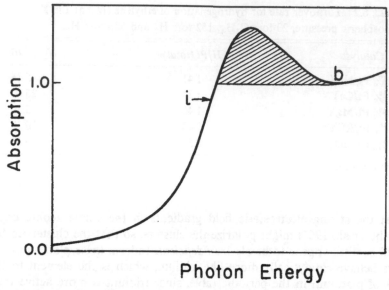

Fig. 6.6 X-ray absorption L_{III} edge: the threshold is represented by its inflection point. The dashed region measures the area of the white line.

the platinum. Similarly, the area of the white line increases as one goes from non-acidic to acidic Pt-zeolites. In both situations, the electron deficiency of platinum in the clusters contained in acidic zeolites is confirmed. In turn, the idea of Schwab concerning a true metal-support interaction with effect on the metal, receives its first chemical and physical confirmation.

What remains unexplained is the nature of the polarization, or charge-transfer complex, in the interaction between the platinum clusters and the framework of the acidic zeolites. All results are collected in Table 6.3.

TABLE 6.3   Clusters of platinum in Y-zeolites

	Non acidic Y	Acidic Y
Size	identical	
Structure	normal	normal[1]
Catalytic activity	normal	enhanced
Chemisorption of oxygen	normal	repressed[2]
Electronic density of Pt	normal	deficient

[1] Gallezot: personal communication
[2] Weber (1982)

Similar metal-support interactions have been reported by Tauster et al. (1978). Because they also are strong, they have been described under the acronym of SMSI. It is suggested here that this useful acronym be used with the meaning: Schwab metal-support interaction, to recognize the pioneering work of Schwab.

6.3 Conclusion

This chapter presents a number of pitfalls to be avoided in the study of chemical, physical, and kinetic phenomena in metal-supported catalysts. Failure to recognize these parasitic effects would make it difficult or impossible to decide whether a reaction is structure-insensitive or not. After taking care of intrusions of heat and mass transfer, there remains to eliminate apparent effects of support. In fact, only one such effect is genuine: the Schwab effect of the second kind. Great interest in this SMSI is due to the possibility of changing catalytic activity and selectivity of metallic clusters by taking advantage of polarization of the latter by the support. Also, catalysts that are more resistant to poisons (sulfur) might be obtained.

In conclusion, if we remember the experimental pitfalls sketched in this chapter, it is even more remarkable that recently so many examples of structure-insensitive reactions have been recognized, and that excellent agreement has been obtained repeatedly between values of turnover frequency on supported clusters devoid of any bulk (100 percent exposed) and on clean, well-defined surfaces of macroscopic single crystals. This agreement is a reflection of the progress in experimental kinetics, through which values of turnover frequency can now be reproduced in different laboratories devoted to surface science and catalytic science.

REFERENCES

Baker, R. T. K. and Sherwood, R. D. 1980. *J. Catal.* 61:378.

Benson, J. E., Kohn, H. W., and Boudart, M. 1966. *J. Catal.* 5:307.

Boudart, M. 1968. *Kinetics of Chemical Processes.* Englewood Cliffs, N. J.: Prentice-Hall, p. 144.

Boudart, M. 1969. *Advan. Catal. Relat. Subj.* 20:153.

Boudart, M. 1979. Paper presented at National Meeting of the American Institute of Chemical Engineers; Abstract No. 77B.

Boudart, M., Vannice, M. A., and Benson, J. E. 1969. *Z. Phys. Chem. N. F.* 64:171.

Briend-Faure, M., Jeanjean, J., Kermarec, M., and Delafosse, D. 1978. *Discuss. Faraday Soc.*, 1538.

Brown, M., Peierls, R. E., and Stern, E. A. 1977. *Phys. Rev.* B15:738.

Carter, J. L., Lucchesi, P. J., Sinfelt, J. H., and Yates, D. J. C. 1965. *Proc. 3rd Intl. Cong. Catalysis*, ed. W. M. H. Sachtler, G. C. A. Schuit, and P. Zwietering. p. 664. Amsterdam: North Holland Publishing Company.

Chambers, R. P. and Boudart, M. 1966. *J. Catal.* 6:141.

Dalla Betta, R. A. and Boudart, M. 1973. *Proc. 5th Intl. Cong. Catalysis*, ed. J. W. Hightower, 2:1329. Amsterdam: North Holland.

Dalla Betta, R. A. and Boudart, M. 1975. *J. Chem. Soc. Faraday Trans.* 72:1723.

Dalmai-Imelik, G., Leclercq, C., Massardier, J., Maubert-Franco, A., and Zalhout, A. 1974. *Proc. 2nd Intl. Conf. on Solid Surfaces, J. Appl. Phys.* Suppl. 2, Pt. 2, 489.

Delgass, W. N. Garten, R. L., and Boudart, M. 1969. *J. Phys. Chem.* 73:2970.

Denbigh, K., and Turner, R. 1970. *Introduction to Chemical Reaction Design.* Cambridge: Cambridge University Press.

Den Otter, G. J. and Dautzenberg, F. M. 1978. *J. Catal.* 53:116.

Dumesic, J. A., Topsøe, H., Khammouma, S., and Boudart, M. 1975. *J. Catal.* 37:503.

Foger, R., and Anderson, J. R. 1978. *J. Catal.* 54:318.

Froment, G. F. and Bischoff, K. B. 1979. *Chemical Reaction Analysis and Design.* New York: John Wiley.

Fuentes, S. and Figueras, F. 1978. *J. Chem. Soc. Faraday Trans I* 74:174.

Fujimoto, K. and Toyoshi, S. 1981, *Proc. 7th Intl. Cong. Catalysis*, ed. T. Seiyama and K. Tanabe, p. 235. Tokyo: Kodansha.

Fujimoto, K. and Boudart, M. 1979. *Journal de Physique* 40:C2-81.

Gallezot, P., Datka, J., Massardier, J., Primet, M., and Imelik, B. 1977. *Proc. 6th Intl. Cong. Catalysis*, ed. G. C. Bond, P. B. Wells, and F. C. Tompkins, 2:696. London: The Chemical Society.

Gallezot, P., Bienenstock, A., and Boudart, M. 1978. *Nouveau J. Chim.* 2:236.

Gallezot, P., Weber, R., Dalla Betta, R. A., and Boudart, M. 1979. *Z. Naturforsch* 34a:40.

Heald, S. M. and Stern, E. A. 1977. *Phys. Rev.* B16:5549.

Holstein, W. L. and Boudart, M. 1981. *J. Catal.* 72:328.

Ioffe, M. S., Kuznetsov, B. N., Ryndin, Y. and Yermakov, Y. 1977. *Proc. 6th Intl. Cong. Catalysis*, ed. G. C. Bond, P. B. Wells, and F. C. Tompkins, 1:131. London: The Chemical Society.

Khoobiar, S. 1964. *J. Phys. Chem.* 68:411.

Khoobiar, S., Peck, R. E., and Reitzer, B. J. 1965. *Proc. 3rd Intl. Cong. Catalysis*, 1:338. Amsterdam: North Holland Pub. Co.

Lacroix, M., Pajonk, G., and Teichner, S. J. 1981. *Proc. 7th Intl. Cong. Catalysis*, ed. T. Seiyama and K. Tanabe, p. 279. Tokyo: Kodansha.

Ladas, S., Dalla Betta, R. A., and Boudart, M. 1978. *J. Catal.* 53:356.

Levy, R. B. and Boudart, M. 1974. *J. Catal.* 32:304.

Lewis, P. H. 1968. *J. Catal.* 11:162.

Madon, R. J. and Boudart, M. 1982. *I & EC Fundamentals* 21:438.

Maurel, R., Leclercq, G., and Barbier, J. 1975. *J. Catal.* 37:324.

Mears, D. E. and Boudart, M. 1966. *AIChE J.* 12:313.

Ménorval, L. C. de, Ito, T., and Fraissard, J. 1982. *J. Chem. Soc. Faraday I* 78:403.

Mestdagh, M. M., Stone, W. E., and Fripiat, J. J. 1972. *J. Phys. Chem.* 76:1220.

Mourgues, L. de, Perrin, M., Sattonay, F., and Trambouze, Y. 1965. *Bull. Soc. Chim. Fr.*, p. 843.

Pashley, D. W. 1956. *Phil. Mag.*, Suppl. 5: 174.

Rabo, J. A., Schomaker, V., and Pickert, P. E. 1965. *Proc. 3rd Intl. Cong Catalysis*, 2:1264. Amsterdam: North Holland Pub.

Schlatter, J. C. and Boudart, M. 1972. *J. Catal.* 25:93.

Schwab, G. M. 1946. *Trans. Faraday Soc.* 42:689.

Schwab, G. M. 1950. *Discuss. Faraday Soc.* 8:166.

Schwab, G. M. 1957. *Naturwiss.* 44:32.

Schwab, G. M., Block, J., and Schultze, D. 1959. *Angew. Chem.* 71:101; *Naturwiss.* 46:13.

Sinfelt J. H. 1979. *Revs. Mod. Phys.* 51:569.

Sinfelt, J. H., Hurwitz, H., and Rohrer, H. C. 1960. *J. Phys. Chem.* 64:892.

Taghavi, M. B., Pajonk, J., and Teichner, S. J. 1978. *Bull. Soc. Chim. Fr.* 1:180.

Tauster, S. J., Fung, S. C., and Garten, R. L. 1978. *J. Am. Chem. Soc.* 100:170.

Taylor, H. 1960. *Proc. 2nd Intl. Cong. Catalysis*, 1:159. Paris: Editions Technip.

Teichner, S. J., Mazabrard, A. R., Pajonk, G., and Hoan-Van, C. 1977. *J. Colloid and Interf. Sci.* 58:88.

Vanhove, D., Makambo, P., and Blanchard, M. 1979. *J. C. S. Chem. Comm.*, p. 605.

Vanhove, D., Makambo, P., and Blanchard, M. 1980. *J. Chem. Research* (M):4119.

Vannice, M. A., Boudart, M., and Fripiat, J. J. 1970. *J. Catal.* 17:359.

Vedrine, J. C., Dufaux, M., Naccache, C., and Imelik, B. 1978. *J. Chem. Soc. Faraday, Trans. I* 74:440.

Weber, R. S. 1982. Ph.D. diss. Stanford University.

Weisz, P. B. 1975. *Ber. Bunsenges* 79:798.

Wong, S. S., Otero-Schipper, P. H., Wachter, W. A., Inoue, Y., Kobayashi, M., Butt, J. B., Burwell, R. L., Jr., and Cohen, J. B. 1980. *J. Catal.* 64:84.

AUTHOR INDEX

SUBJECT INDEX

Library of Congress Cataloging in Publication Data

Boudart, Michel.
Kinetics of heterogeneous catalytic reactions.

Translation of: Cinétique des réactions en catalyse hétérogène.
Includes bibliographical references and indexes.
1. Heterogeneous catalysis.   2. Chemical reaction, Rate of.
I. Djéga-Mariadassou, G.   II. Title.
QD505.B6913   1984   541.3'95   83-43062
ISBN 0-691-08346-0   ISBN 0-691-08347-9 (pbk.)

Milton Keynes UK
Ingram Content Group UK Ltd.
UKHW010840140924
448309UK00008B/326

9 780691 612560